W9-CQJ-954

Climate's Impact on Food Supplies

Strategies and Technologies for Climate-Defensive Food Production

AAAS Selected Symposia Series

Published by Westview Press, Inc.
5500 Central Avenue, Boulder, Colorado

for the

American Association for the Advancement of Science
1776 Massachusetts Avenue, N.W., Washington, D.C.

Climate's Impact on Food Supplies

Strategies and Technologies for Climate-Defensive Food Production

Edited by
Lloyd E. Slater and Susan K. Levin

AAAS Selected Symposium **62**

AAAS Selected Symposia Series

This book is based on a symposium which was held at the 1980 AAAS National Annual Meeting in San Francisco, California, January 3-8. The symposium was sponsored by the AAAS Climate Project.

Published in 1981 in the United States of America by
Westview Press, Inc.
5500 Central Avenue
Boulder, Colorado 80301
Frederick A. Praeger, Publisher

Library of Congress Cataloging in Publication Data
Main entry under title:
Climate's impact on food supplies.
(AAAS selected symposium ; 62)
Includes bibliographical references.
1. Crops and climate--Congresses. 2. Meteorology, Agricultural--Congresses. 3. Food supply--Congresses. 4. Agriculture--Congresses. I. Slater, Lloyd E. II. Levin, Susan K. III. Series.
S600.2.C56 338.1'9 81-753
ISBN 0-86531-166-8 AACR2

Printed and bound in the United States of America

About the Book

Global concern about the impact of climatic variability on food supplies has been growing since 1963 when a major crop failure in the Soviet Union ended a long period of abundant, low-cost surplus grain. Since then a number of sizable crop shortfalls and food supply emergencies throughout the world have encouraged the study of climate-food interactions--a complex undertaking. This book assesses theories of how climate influences food production and distribution, discusses how societies can contend with climate-induced food shortfalls, and explores policies and technologies that promise to reduce climate's impact on agriculture and to increase food production through unorthodox methods.

About the Series

The *AAAS Selected Symposia Series* was begun in 1977 to provide a means for more permanently recording and more widely disseminating some of the valuable material which is discussed at the AAAS Annual National Meetings. The volumes in this *Series* are based on symposia held at the Meetings which address topics of current and continuing significance, both within and among the sciences, and in the areas in which science and technology impact on public policy. The *Series* format is designed to provide for rapid dissemination of information, so the papers are not typeset but are reproduced directly from the camera-copy submitted by the authors. The papers are organized and edited by the symposium arrangers who then become the editors of the various volumes. Most papers published in this *Series* are original contributions which have not been previously published, although in some cases additional papers from other sources have been added by an editor to provide a more comprehensive view of a particular topic. Symposia may be reports of new research or reviews of established work, particularly work of an interdisciplinary nature, since the AAAS Annual Meetings typically embrace the full range of the sciences and their societal implications.

WILLIAM D. CAREY
Executive Officer
American Association for the Advancement of Science

Contents

About the Editors and Authors

Lloyd E. Slater *is manager of the Food and Climate Forum, an activity of the Aspen Institute. The former editor and publisher of* Food Engineering International *and executive director of Puerto Rico's Institute on Social Technology, he has been directly involved with problems of developing countries and their food supplies. He is the author of numerous publications in food engineering, automatic control technology, and biomedical engineering.*

Susan K. Levin *served as editor of the* 1979 Food and Climate Review, *an annual publication of the Aspen Institute's Food and Climate Forum. Formerly a documentary producer for public radio, she is a part-time farmer but continues to produce reports on food-related topics for radio and Colorado newspapers.*

Martin E. Abel *is senior vice president of Schnittker Associates, a Washington, D.C.-based agricultural economics consulting firm. A specialist in agricultural policy, commodity analysis, and trade, he is the author of* Reducing the Climatic Vulnerability of Food Supplies in Developing Countries: Public and Private Alternatives *(report by Schnittker Associates for the Aspen Institute, 1978).*

Diane C. Brown *is a senior associate for research at Schnittker Associates in Washington, D.C. Her emphasis has been on food supplies, developing countries, and international trade, and she has contributed papers to the President's Commission on World Hunger.*

F. Kenneth Hare *is professor of geography at the University of Toronto and provost of Trinity College. His specialties are climatology and desertification and he has published extensively in his field.*

David P. Harmon, Jr., *director of the Food Advisory Board, has focused his research on the future of agriculture. He is coeditor of* Critical Food Issues of the Eighties *(with M. Chou; Pergamon, 1979) and coauthor of* World Food Prospects and Agriculture Potential *(with M. Chou, H. Kahn, and S. Wittwer; Praeger, 1977).*

Carl N. Hodges, *director of the Environmental Research Laboratory at the University of Arizona, is investigating technologies to produce food in hostile environments. His work with salt-tolerant crops and controlled-environment aquaculture/crop systems has attracted worldwide attention.*

James D. McQuigg *is a consulting meteorologist for industry and government and private institutions. Formerly director of the NOAA/EDS Center for Climatic and Environmental Assessment, he has published widely on agrometeorology and climatology.*

Roger R. Revelle, *professor of science and public policy at the University of California at San Diego, was formerly director of the Harvard University Center for Population Studies. He has published numerous scientific articles in his fields of interest.*

Walter Orr Roberts *is director of the Aspen Institute for Humanistic Studies Program on Food, Climate and the World's Future. A solar astronomer, he was the first director of the National Center for Atmospheric Research. In recent years he has devoted much of his time to the problem of climate change and the interplay between climate and food; his most recent book is* The Climate Mandate *(with H. Lansford; W.H. Freeman, 1979).*

Norman J. Rosenberg *is professor and director of the Center for Agricultural Meteorology and Climatology at the University of Nebraska-Lincoln. Currently he serves as a consultant to USAID and NOAA on problems of crop production in semi-arid regions. The author of numerous technical articles, he is the editor of* North American Droughts *(AAAS Selected Symposium 15; Westview, 1978) and* Drought in the Great Plains: Research on Impacts and Strategies *(Littleton, Colorado: Water Resources Publications, 1980).*

Stanley Ruttenberg *is secretary general of the International Association of Meteorology and Atmospheric Physics. Based at the National Center for Atmospheric Research, he has published widely on advanced observing systems for meteorological and climatological research and various aspects of atmospheric physics.*

John A. Schnittker, *president of Schnittker Associates in Washington, D.C., was undersecretary of agriculture from 1965-1969. He specializes in U.S. and world food and agricultural policy.*

Howard E. Worne, *president of Worne Biochemicals, Inc., is a pioneer in automated soil and tissue analysis. He is the author of numerous publications on industrial microbiology and biochemistry and has served as a consultant on biological agriculture for various governments and Indian tribes.*

Acknowledgments

The editors wish to express their appreciation to the Aspen Institute for Humanistic Studies and to the sponsors of the Institute's Food and Climate Forum for offering the time and considerable supporting services required in preparing this book. They are grateful also to Jo Ann Green for "polishing" and typing much of the material and to Dr. Kathryn Wolff and Joellen Fritsche of AAAS for their encouragement in making the book a reality.

Climate's Impact on Food Supplies

Strategies and Technologies for Climate-Defensive Food Production

Walter Orr Roberts

Introduction

From the earliest times, scholars have speculated about the waxing and waning of civilizations. In his *Timaeus*, written around 400 B.C., Plato recounts the advice given to Solon, the wisest of the Seven Sages, by an aged Egyptian priest. "Solon, Solon, you Greeks are eternal children," the priest warned. "There have been many occasions of human destruction, and there will be many more. The chiefest source of these is fire and water..." By the words fire and water, the priest meant the scourges of drought and flood that had ravaged Greece. Plato's priest went on to say that the Egyptians, unlike the Greeks, preserved ancient lore and history. "Whenever anything great or glorious or otherwise noteworthy occurs, it is written down and preserved in our temples; whereas among you and other nations,...recently endowed with the art of writing and civilized needs,...there has recurred like a plague brought down upon you a celestial current (meaning the droughts and floods), leaving only an unlettered and uncivilized remnant; wherefore you have to begin all over again like children, without knowledge of what has taken place in older times either in our land or yours...."

Rhys Carpenter, writing in 1965, used these and other recollections from Plato to preface his fascinating book, *Discontinuity in Greek Civilization*. It is difficult for us today to think of ancient Greece as a developing country compared to other nations of its time, but it is engrossing to realize there is so old a perception of climate as a major cause of human disaster.

In the same book, Carpenter speculates that the fall of the Greek city Mycenae, around 1200 B.C., was due to successive years of severe drought, even though Athens, less than 100 km away, remained well-watered. In Mycenae, food riots and violent uprisings by the poor against authority led to the city's downfall. This may be one example of the kind of human destruction the aged priest was telling Solon about.

The Egyptian priest's caveat to Solon, though centuries-old, nonetheless still holds: as long as the celestial currents bring down upon us devastating droughts and floods, man must take notice and dutifully record these occurrences, lest he "begin all over again like children."

20th Century Explanation

It awaited Reid Bryson of the University of Wisconsin to find a plausible meteorological explanation for how such a drought might have occurred in Mycenae, leaving nearby Athens unscathed. In recent technical papers and in an excellent popular book, *Climates of Hunger*, Bryson suggests, with due caution, that a large-scale weather circulation pattern showing up in highly-reliable, modern records might provide the answer. The weather pattern would simply have had to become relatively more frequent and persistent to produce the juxtaposition of drought in Mycenae, yet ample rainfall in Athens.

Other popular books by authors such as Ellsworth Huntington, Nels Winkless III and Iben Browning also call attention to recurring significant impacts of climate fluctuations and trends on human affairs. One of the best recent books on the subject is *The Genesis Strategy* by Stephen H. Schneider. Schneider graphically discusses the dependence of human well-being on stable climate, and calls for a worldwide political strategy, a "genesis strategy," to establish international grain reserves large enough to accommodate human crises triggered by climate, wherever and whenever they occur -- as most certainly they will. His strategy, based on the story in the Book of Genesis in which Joseph interprets the Egyptian Pharaoh's dream, is to store grain in the "seven good years" for the "seven lean years."

Though millenia have passed since the fall of Mycenae, it was not really until the year 1972 that public attention everywhere focused on the critical equation of food, climate and human welfare. In that year, several climate anomalies in different geographical locations had serious consequences: 1) an extraordinarily hot summer and very little rain in the Moscow area caused the Soviets to resort to massive grain purchases on the international market to maintain livestock herds; 2) The Indian monsoon commenced late, was interrupted in mid-season, and ended early, thereby heavily damaging grain crops; 3) the six-year drought in the Sahel region of Africa climaxed with near total eradication of cattle herds of the nomadic peoples of the region; and 4) the anchoveta crop of Peru suffered a calamitous failure, wreaking havoc with fisheries and materially reducing the protein supplies of Latin America. Total world production of food downturned for the first time in many years, and though the aggregate decline was but two percent, the impact in some localities approached disaster.

The pitiful state of refugee nomads of the Sahel saturated the press everywhere, and the impact of the African drought, which extended all the way to Ethiopia, became a driving factor in planning the United Nations Conference on Desertification. It built momentum for the World Food Congress in Rome in 1974, and gave great thrust to a drive to establish world grain reserves for just such disasters. In the United States, it spurred the entirely voluntary Hunger Project, with its aim to end world hunger within a decade. Heightened consciousness of the growing specter of malnutrition among the world's poor spawned the Presidential Commission on World Hunger.

Research Sharpens the View

In May of 1974, the International Federation of Institutes for Advanced Study (IFIAS) assembled an international planning workshop at the University of Bonn to identify top-priority studies on climate's impact on society, which merited research of a cross-disciplinary nature. From among over 20 suggested topics, attention focused on the task of examining the social, political, economic and ethical impacts of climate changes on the character and quality of human life.

Within a few months, and before a detailed plan of the IFIAS study had materialized, the magnitude of the 1972 anomalies and the severity of their human implications reached the attention of all. In the spring of 1975, the IFIAS project firmed up around the notion of examining specifically the topic of "Drought and Man: The 1972 Case." And in the fall of 1974, IFIAS issued a widely circulated appeal for new attention to the problem. Excerpts on changing climate from the 1974 meeting statement still have a strong ring today:

> "We must anticipate that such deviations or 'anomalies' will recur. At this moment the world is unprepared to cope with them. Grain reserves which used to be abundant in some regions are no longer sufficient to serve as insurance against disaster and by some estimates have dropped to

such low levels that they can supply the world needs for less than one month at present consumption rates. At the same time wasteful and excessive consumption by the affluent, along with increasing numbers of mouths to feed, strains the capacity of farmers to deliver enough food even from the best of harvests. It becomes ever more difficult, expensive and risky to open up new arable land, and at least as difficult to limit the use of marginal lands highly vulnerable to erosion and worsening of climate.

"In short, the current food-production system now has little flexibility with which to meet emergencies. What we have hitherto regarded as occasional emergencies, moreover, can no longer rationally be so regarded.

"The nature of climate change is such that even the most optimistic experts assign a substantial probability of major crop failures within a decade. If national and international policies do not take such failures into account, they may result in mass deaths by starvation and perhaps in anarchy and violence that could exact a still more terrible toll. It would be irresponsible in these circumstances to continue passively in our present condition of helplessness: without food reserves or alternative technologies to produce food, and without adequate means to redistribute food from the more favored nations or more favored groups within nations to the less favored in time of urgent need."

The distinguished Argentine scholar Rolando V. Garcia was selected to head the IFIAS three-year study effort. From this investigation has come a new perspective on food-climate interactions. The widespread publicity of the misery of the tragically malnourished children of regions of Africa and in Bangladesh resulted in a perhaps overly-simplified perception by most people of climate's impact on food supplies. The report of the IFIAS study under Garcia draws a full picture of the multifarious components surrounding and incorporating the food-climate connection. I strongly recommend Garcia's volume. It has in it some harsh judgements about where blame lies, and some will contest a number of its imputations. But it is an important step towards the development of a more balanced perspective about the root causes of world hunger. Only if new and more balanced perspectives become widespread will we have much hope, I suspect, of conquering malnutrition among the world's poor.

While another article in this volume gives some of the principal findings of Garcia's IFIAS study, it is enough for me to say here that the human tragedy resulting from a drought or a flood is not simply the food lost. It is a far more complex socio-political matter, somewhat exemplified by this passage from the *Politics of Starvation* by Jack Shepherd and published by the Carnegie Endowment for International Peace:

> "Between March and September, 1973 -- the worst periods of the famine in Wollo and Tigre provinces -- there were some 20,000-30,000 tons of grain stored in commercial warehouses around Ethiopia....An American embassy officer in Ethiopia reported seeing 'peasants starving to death within a few miles of grain storage'....As long as large amounts of relief grain were held off the market or, preferably, not even brought into the country, shortages (and high prices) could occur and profits could be made.... As prices rose, at one point the Ethiopian government offered to sell 4,000 metric tons of grain it had in storage to the United States, which could then donate it back for relief inside Ethiopia."

The Problem Intensifies

Today nearly one-fourth of humanity lives on the razor edge of damaging malnutrition. These are the world's poorest, and perhaps a half of them will lead short and tragic lives as a consequence of their plight. The problem is not going away as world affluence grows; instead, it seems destined to intensify year by year. By the end of the next five or six decades I expect we will see our world characterized by:

- doubled world population,
- tripled world food demands,
- quadrupled world energy consumption.

With these changes will come mounting costs for pollution control; worsening shortages of good quality water for irrigating the 17 percent of the world's cultivated land that now yields 50 percent of all agricultural products; accelerating degradation of the environment as more marginal lands are farmed and as world forests are decimated; increasingly expensive food in world markets -- and the list could go on much further. Moreover, it is entirely possible that meeting these demands will produce unplanned human impacts on the stability and character of the earth's climate.

Most climate experts are convinced, for example, that carbon dioxide in the atmosphere will double over this time span, mainly as a result of the burning of fossil fuels for energy use. With this CO_2 doubling we will suffer an average climate warming of two to three degrees C in mid-latitudes. This is a change of magnitude probably unprecedented since *Homo sapiens* first walked the earth.

If the anticipated CO_2-induced warming happens, there will be considerable effects on the world's agriculture. Some regions of the earth are expected to have more rainfall, some less. On balance, such a climate change could do more harm to world food supplies than good. With more mouths to feed, we will probably have a difficult time producing the needed food from most of the earth's present 1.5 billion hectares of cultivated land.

We clearly face a time of challenge and of tough choices. I am confident that innovative approaches will make it possible to meet human needs, but I am also certain that in so doing we will have to develop new perspectives not only on food and climate, but also on mankind's way of doing things politically, socially, economically and morally. It is not a time for business as usual, if we hope to see business survive at all.

Part 1

Viewing the Climate/Food Interaction

F. Kenneth Hare

1. Climate's Impact on Food Supplies: Can It Be Identified?

Introduction

The serious study of climatic impact on human society was a phenomenon of the 1970s -- as it was in the 1930s. In those two decades climate upset the economic apple-cart severely enough to be noticed on the world scale. We are now into a new decade. Will the mood last much longer?

Concern will certainly remain alive if this emerging decade produces the same scale of climatic disturbance as the 1970s. We have no means of predicting such an extension, but it may well happen. And concern may remain alive if the scientific and technical professions learn how to characterize the impact...and how to recommend remedial action. Otherwise politicians and statesmen will give the notion of understanding and dealing with climatic impact short shrift, even if the disturbances continue. They have little use for insoluble problems that the media ignore. Already the energy crisis and Middle Eastern politics have chased climatic impact off the front pages. Unless we professionals can show how climatic impact should be handled -- and, if possible, predicted -- it will not compete for world attention.

Old enough to have lived through both decades, I was highly conscious of each of them, having opted for climatology at the age of four and persisted with it over the years. The 1930s, with their widespread drought, are still most vividly with me. They were years of vast crop and soil losses and desperate human suffering in several parts of the temperate world. It was the decade of the Dust Bowl, of the Grapes of Wrath, and of the great depression. Moreover, it was a time before society had put in place strong institutions with the skills and resources to combat the effects. The New Deal in the United States itself owed much to the American perception of the drought and soil erosion. Of course the

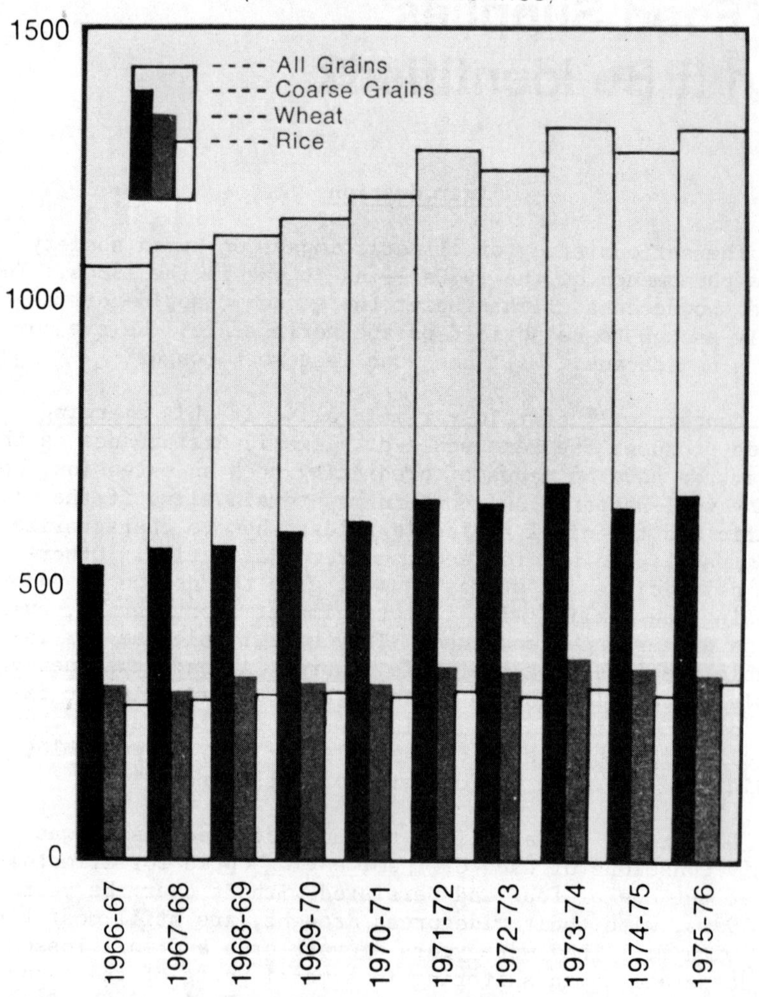

Fig. 1. World grain production 1966 – 76. Courtesy of the Canadian Wheat Board.

depression was not caused by drought, yet in many ways the depression itself compounded the effects of climatic stresses. Nevertheless, the losses of soil and distress of the dirt farmer were both a prime target of New Deal measures, and a highly visible testimony to the power of climate to disrupt human affairs.

Perception Still Lacking

Even with these vivid previous experiences, and in spite of the efforts of pioneers like Walter Orr Roberts, Rolando Garcia and Michael Glantz, we are entering the 1980s with a very incomplete perception of climatic impact. This is a serious matter, since the United Nations system is now committed to a World Climate Program in which the identification of such impact is one of three major components. Also, the United States and Canada have comparable national programs. Yet, it is proving difficult for all concerned to come to grips with the impact problem.

Without question the major failing has been our inability to specify the impact of climate on the largest economic scales -- those of international trade, of the commodity markets and of national economic policies. At such levels the impact is necessarily delayed. When it finally arrives, it is inextricably mingled with other influences, some of which are related to climatic factors, and some of which are not. It may not be feasible to seek separate identifications of climatic impact on such scales.

In any case, climatic anomalies tend to cancel one another when they are averaged over sufficient space. Drought in one large continental area will generally be counterbalanced by good growing conditions in another. Throughout the 1950s and 1960s, for example, world production of cereals rose fairly steadily, as shown in Fig. 1, even though there were significant crop failures in several years over major producing areas. As Fig. 2 indicates, the record of production from a single area is much less stable. Differences in cereal output between years from producers as large as the Soviet Union may exceed 10 percent on either side of the long-term trend. Swaminathan (1979) has claimed some positive correlation between crop production anomalies over the world as a whole, but this effect has been small in relation to the general upward trend of production.

In the 1970s, however, the upward trend faltered, for a wide variety of reasons. The crop failures of 1972-73, seen in Fig. 3, were associated with widespread climatic anomalies of several kinds. Major producers of wheat were severely

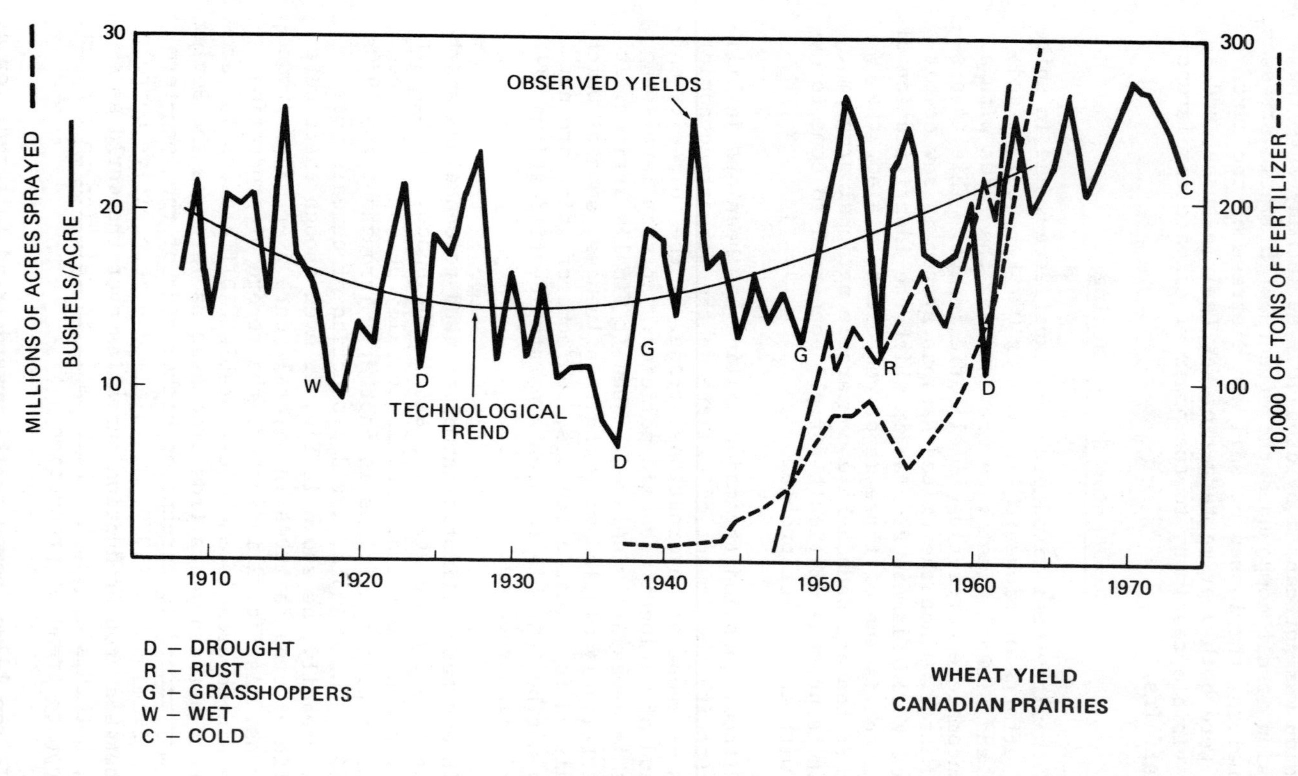

Fig. 2. Wheat yield per unit acre, as function of crop year and treatment. Courtesy G. McKay, 1976.

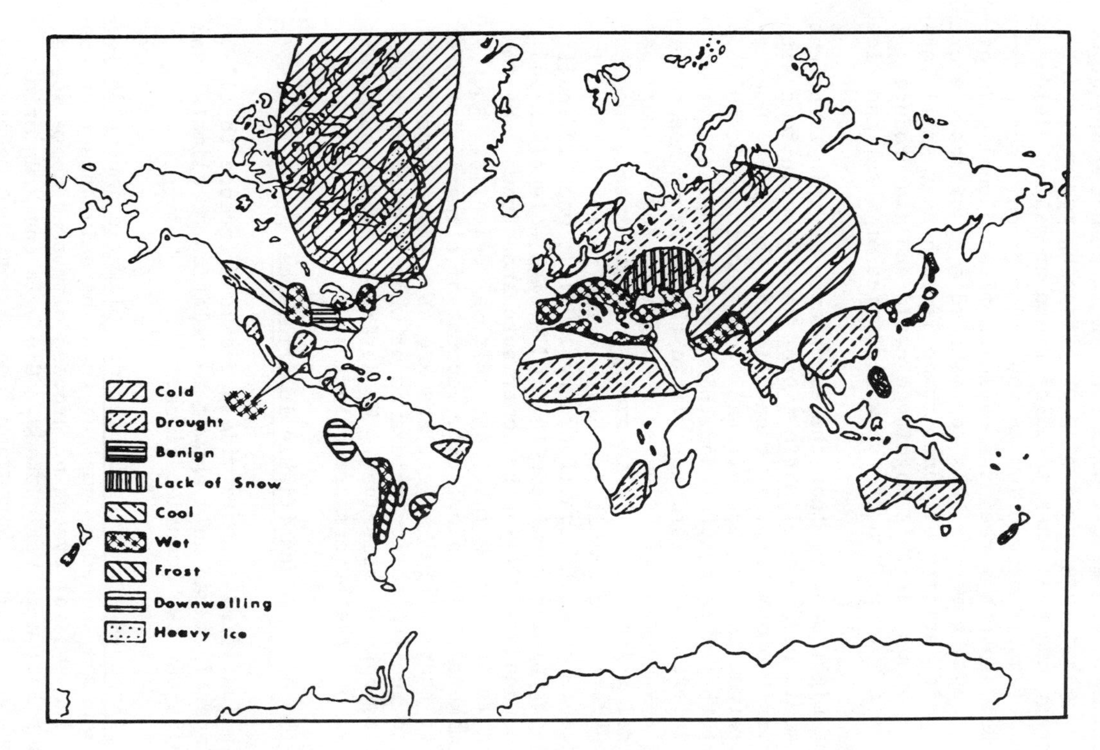

Fig. 3. Climatic anomalies, 1972. Courtesy G. McKay (Personal communication)

affected by drought or winter-kill, notably in the Soviet Union. The market disturbance due to panic buying was profound, and ushered in several years of price inflation and disrupted storage. Subsequent years in the decade produced large climatic anomalies, with crop failure in the Soviet Union and other large consuming or producing regions. Yet the largest and most unexpected of these anomalies yielded a downturn of world production on the order of only one percent. Set against a recent experience of an annual gain of three percent, however, this represented a significant shortfall. The elasticity of world trade is such that small variations of supply of this magnitude could and did produce doublings and triplings of the world price of wheat and rice.

One might suppose that the experience of this turbulent decade would have put climatic impact on the world food situation firmly on the map. But such is not the case. One recent international agenda virtually omitted it (FAO, 1979). The fact that this chapter and this book are being written is due to interest and awareness within the American Association for the Advancement of Science, a congeries for the most part of natural scientists. It is among such as us that one finds consciousness of climatic impact. At the recent World Climate Conference in Geneva (February 1979) it was difficult for the organizing committee to find social scientists and economic managers willing to contribute and attend. My impression is that economists, agricultural scientists and policy-makers still take little account of climatic factors. I suspect that natural scientists are as a whole more inclined to see the policy relevance of what they do than are the policy-makers themselves. We desire to export our convictions, but the buyers are reluctant.

A Justifiable Indifference?

How justified is this widespread indifference to climatic impact? I can only answer in relation to climate, and must point out two considerations:

1) Impacts are often buffered. Climatic impact on food production may indeed be smaller than the naive observer supposes. The spatial averaging process discussed above works well in relation to world trade. Gluts in one area offset failures in another. Moreover, feedbacks galore modify the direct impact of climate within all the advanced countries. Central marketing agencies, government subsidy programs, trade and professional associations and even the media influence acreages and crop varieties planted, as well as spot and future prices. These processes are sufficiently prompt in their operation to build in significant countermeasures as

soon as a climatic anomaly is perceived. In my own country, Canada, a year of poor harvests in the Prairie Provinces does not necessarily reduce overall economic production, or even farm incomes. The response of the food system to external, natural stimuli is hence strongly non-linear and lagged in time, and it will become more so.

2) <u>But less so in LDCs.</u> Clearly the arguments advanced above apply primarily to the advanced countries; they have little force in the LDCs. The impact of the Sahelian drought was singularly direct and uncushioned for thousands of nomads and subsistence cultivators near the desert margin. Even here, however, there were non-linearities that make a simple identification of climatic impact misleading (Kates, Johnson and Johnson Haring, 1977; Garcia, 1978; Oguntoyinbo and Odingo, 1979). The introduction of veterinary medicine, the growth in human population, and post-colonial policies regarding land-use and international migration affected the precise nature of the impact. It is very probable that the ancient systems of nomadic pastoralism could have survived the 1968-73 drought more readily than did the modified social systems of today. But the latter accommodate many more people, and it is not at all clear that the older tradition could have coped with modern population numbers.

Taking these two considerations together, it is clear that climatic impact is a misnomer. One must talk instead, of <u>climate-society interactions</u>. Even simple societies respond quickly to the perception of climatic stress, and put in place counter-measures that minimize the effects. And on the small scale such measures may significantly affect the microclimate itself; a field left uncultivated under protective stubble has a slightly different thermal climate from one where soil is allowed to lie naked. A recent 1979 SCOPE Workshop attempted to depict this interactive process in a formal structure, as shown in Fig. 4.

Reviewing the Linkages

Referring to the tabulation in Fig. 4, much evidence was taken on its five points at the World Climate Conference, which reviewed the climate-economy link in each main sector of the world economy. As might be expected, a bewilderingly complex set of interactions emerged. Nevertheless, certain general conclusions can be drawn from the published papers, which I had the dubious pleasure of convening (WMO, 1979):

1) <u>Has climate changed</u>? It was agreed that the disruptions of the world food system during the 1970s were due mainly, if not entirely, to natural short-term variability of

No.	Level of Impact	Disciplines Relevant to Analysis	Change Occurring in
1	Climatic parameters	Climatology, Oceanography	Temperature, rainfall, radiation, etc.
2	Potential related environmental change	Ecology, Physical Geography	Ecosystem and habitat changes, etc.
3	Human activity	Economics, Sociology, Psychology, Anthropology, Geography	Agriculture, energy use, tourism, transport, construction, manufacturing, etc.
		-----PERCEPTION-----	
4	Social response and interaction	Sociology, Psychology, Geography, Anthropology	Employment level, migration, culture, social conflict, etc.
5	Political processes and responses	Economics, Political Science, Law	Legislation and regulation, legal action, research and development programs, international action and policies.

Fig. 4. The Climate-Society Interaction

climate, i.e. to fluctuations or extremes within the frequency-spectrum of existing climate. Such variability may be expected to produce interannual variations of world grain yield of greater than 21 million tons one year in three (USDA, 1975). Even the 1968-73 Sahelian drought was seen by most observers as a fluctuation natural to the existing climate. No convincing evidence was brought forward of a permanent shift toward a climate more hostile to world food production -- though it was equally true that no proof of the contrary was forthcoming (Hare, 1979; Mason, 1979).

2) Information poorly used. It was generally agreed that economic strategies in all the main sectors, especially energy, agriculture and fisheries, made poor use of existing information about such short-term variability. There was a large perceived need for better use of data already on file, and for better understanding of climate's direct impact on each sector's performance. The World, U.S. and Canadian climate programs aim to achieve this improvement.

3) Predictability is required. Equally, however, there was a conviction that climatic information can have little overall effect on national and international policies unless it becomes predictive. Much of the delegates' time was taken up in discussions of predictability, and of the research needed to improve the present situation -- namely, that variability on the climatic time-scale (i.e. beyond a few weeks) is wholly unpredictable. Since climatic prediction, if attained, will rest on different concepts from those of numerical weather-prediction, the failure of the latter to achieve skill beyond a few days still leaves open the hope that climatic prediction on the world-scale may be possible. The few economists present in Geneva agreed that such prediction, not achievable for any other economic influence, would have dramatic market and policy effects. But they did not expect the meteorologists to succeed. Economists are not so much dismal, as cynical. Their own failures make them skeptical of other forecasters.

4) When change does occur. Longer-term climatic change due to such influences as carbon-dioxide increase will probably begin to affect food production systems before the end of the century. If rises of world temperature on the scale of three degrees Celsius later occur, and if there are large associated changes of precipitation, the effects are likely to be drastic, and at present unpredictable. Margolis (1979) pointed out an obvious, yet overlooked point. Agriculture and pastoralism must live through the usual and normal brief climatic fluctuation with existing technology. There simply isn't time to develop new varieties of crops and animals, or

to build new irrigation systems. The pace of true climatic change, however, is slow enough to allow such adaptations. If carbon dioxide increase really does transform world climate during the first third of the twenty-first century, there should be ample time to produce the new varieties, the new public works and the new socio-economic institutions necessary to make the transformation -- if, and only if, the reality of the coming change of climate is understood by the world's political leaders.

A Broader Perspective

There has been a tendency for us to see the whole question in narrow sectorial and technical terms. It is worth taking the trouble to attempt a broader outlook. Let me summarize what I myself see as the essential components of the problem -- influenced, I must admit, by the excellent study of Biswas and Biswas (1979):

1) Climate impact is most obvious, and most readily comprehended, in terms of annual crop production in the world's granaries, especially as regards corn, wheat and rice. There is extensive knowledge of the effects of rainfall and temperature anomalies on corn and wheat yield. There is some, but not enough, interaction between plant and animal geneticists and climatologists, and between agricultural meteorologists and crop-production specialists. There is a partially-effective flow of information from weather and climate offices to the individual farmer, mainly in the developed countries. This entire process has been the traditional way of treating climate and food interaction.

2) Much less obvious are the changes induced in the productivity of the system that are traceable to climate, at least in part. These include the losses in soil quality and quantity that have been in progress ever since farming began, but which have been accelerating in the middle and late twentieth century. A recent estimate by Buringh (1979) suggests that almost one-third of the original carbon of the world's soils (2,014 x 10^9 tons) has been lost through oxidation of soil humus. He estimates that the present net loss of soil carbon due to agriculture and forest clearance may be 4.6 x 10^9 tons per annum. Even if this estimate proves too high (and it is much higher than, for example, estimates by Bohn (1976) and Wong (1978), it points to a significant climate-food linkage; carbon lost to the soil is, of course, acquired by the atmosphere, and contributes to the growing greenhouse effect.

3) The most damaging and widely-publicized impact

of climate has been upon the dry margins of rain-fed agriculture and pastoralism. The U.N. Conference on Desertification in 1977 showed that soil erosion, loss of woody and perennial vegetation (including firewood), falls in groundwater tables, and the spread of salinization in irrigated areas, were imperiling the economic base of many tropical countries, and were by no means negligible in the Soviet Union, China, the United States and Australia. The verdict of the Conference (UNCOD,1977) was that socioeconomic maladjustment to environment was the direct cause of these losses, but that climatic stress, especially prolonged drought, accelerated the decline. Much of the attention of the world's major institutions -- the World Bank, the foundation-led Green Revolution and the governments of industrial nations -- had been given to the problem of intensive agriculture of irrigated lands in the humid tropics. The most vulnerable lands and people, it was argued, might well lie in the vast areas of rain-fed agriculture in the drier tropical countries. This conviction has already led to a significant change of stress in the programs of the international crop research institutes, but the impression remains that the productivity of vast areas of the dry tropical world is being irreversibly diminished...and with it, the livelihood of innumerable subsistence cultivators and pastoralists. Debate about the world food system often focuses on the international grain trade, and on prices in advanced-country supermarkets. For the inhabitants of Niger, Haute Volta, Chad, Ethiopia and Somalia such debate is empty and meaningless.

Summary

In this short review of the interactions between climate and the world's food systems, I have concentrated on the question of <u>identifying</u> the impact of climate. Such identification is essential if counter-measures are to be taken. Inevitably, I have stressed the problems perceived in the past decade, and have been much swayed by my own experience -- essentially as a researcher, author, and editor in connection with two U.N. conferences on the general theme. Hence, I must close by warning the reader that my view is certainly incomplete, and perhaps quite wrong. I invite all readers concerned with the issue of climatic impact to admit as much about their own limitations. And I stress once again that authorities outside the atmospheric sciences tend to see this problem in starkly different terms. Sometimes they even deny its existence. At the World Climate Conference, M. Swaminathan of the Indian Council of Agricultural Research used these words:

"Given appropriate political decisions and resource

> back-up, this task (providing national and international food security systems) can be accomplished by 1984, the target year set by the World Food Conference...for banishing hunger from the Earth." (Swaminathan, 1979)

Coming as they do from one of the most respected figures in world agricultural science, these remarks should remind us that the climatic factor, though without question important, is not seen as central by large numbers of our contemporaries. My hope is that the discussions following the Seminar in San Francisco which led to this book, and the book itself, will help us toward a balanced judgement.

References

Biswas, M.R. and A.K. Biswas 1979. Food, Climate and Man. Wiley-Interscience, New York.

Bohn, H.L. 1976. Estimate of organic carbon in world soils. Soil Science Society of America Journal, 40, 486-490.

Buringh, P. 1979. Decline of organic carbon in soils of the world. Center for World Food Studies, Agricultural University of the Netherlands, Wageningen. (to be published)

Canadian Wheat Board 1979. MS Statistics of the World Grain Trade.

Food and Agricultural Organization 1979. Background paper for discussion purposes, MS.

Garcia, R. 1978. Climate impacts and socioeconomic conditions. Annex C in International Perspectives on the Study of Climate and Society, National Academy of Sciences, Washington, D.C.

Hare, F.K. 1979. Climatic variation and variability. In WMO 1979, see below, 51-87.

Kates, R.W., D.L. Johnson and K. Johnson Haring 1977. Population, society and desertification. In UNCOD 1977, see below, 261-317.

Margolis, H.W. 1979. Contribution to SCOPE, 1979, see below.

Mason, B.J. 1979. Some results of climatic experiments with numerical models. In WMO, 1979, see below, 210-242.

McKay, G.A. 1976. Future climate and decision-making. In

R.J. Kopec, ed. Atmospheric Quality and Climatic Change, Studies in Geography No. 8, University of North Carolina, Chapel Hill, N.C. 90-95.

Oguntoyinbo, J.A. and R.S. Odingo 1979. Climatic variability and land use: an African perspective. In WMO 1979, see below 552-580.

SCOPE 1979. Report on ICSU/SCOPE Workshop on Climate/Society Interface. Canadian Climate Center, Atmospheric Environment Service and Institute for Environmental Studies. 36 pp.

Swaminathan, M.S. 1979. Global aspects of food production. In WMO, 1979, 369-405.

UNCOD 1977. Desertification, its Causes and Consequences. U.N. Conference on Desertification Secretariat, Nairobi.

USDA 1975. The world food situation and prospects to 1985. Foreign Agricultural Economic Reports, no.98. United States Department of Agriculture, Washington, D.C.

WMO 1979. Proceedings of the World Climate Conference. World Meteorological Organization, Geneva.

Wong, C.S. 1978. Atmospheric impact of carbon dioxide from burning wood. Science, 200, 197-200.

Stanley Ruttenberg

2. Climate, Food and Society

Introduction

In this chapter I will review all too briefly R.V. Garcia's project on the 1972 drought. It was my inestimable pleasure and honor to have had the opportunity to collaborate -- a junior collaboration to be sure -- with Rolando V. Garcia in his far-ranging and detailed studies of the physical and human aspects of the so-called 1972 food crisis brought on by droughts in many food-producing regions.

The main body of this chapter is a summary of material taken from a draft of Volume I, in which Garcia's main theses are presented, along with syntheses of many collaborative reports that appear in full in two other volumes of the project report. In this endeavor to extract some of Garcia's key findings and rephrase them to preserve the meaning and the context, I am indebted to Susan Levin, Aspen Institute for Humanistic Studies, Food and Climate Forum, Boulder, without whose help and penetrating understanding of the work this summary would not be available.

A summary of this kind cannot do justice to all the material Garcia gathered and the information and "facts" he used to adduce his conclusions. The reader is urged to read Garcia's full report. If he or she wishes to take issue with his interpretations, it should be done on the basis of the whole report, and not on this necessarily limited and selective summary. One should note that I wrote "facts," not facts. Garcia, in the report, makes a most interesting argument about the nature of fact, essentially its non-absolute nature. In Garcia's philosophy, a fact is an abstraction, a construct based on an observable in the context of the interpreter's own experience and intellectual ideology. Garcia makes much use of this distinction. He feels that many observables of the period centered on 1972 can be interpreted in a variety of ways, depending on the context, and even the ideological point of view of the interpreter, to arrive at "facts" that differ from one interpreter to another. He devotes much discussion to the "facts" as set out by the conventional disaster wisdom of the time, as contrasted with his alternative interpretations, partly as an existence of alternative explanations.

The word "ideology" is a subject for some comment here. Garcia himself comes from the developing world, Argentina. He has also lived elsewhere, and has noted the difference in opportunities (economic, social and political) available elsewhere. He reveals himself, at least to me, in his writings and in particular in this report, as a humanist in nonpolitical ideological terms. He has a deep faith that the human reasoning power will, ultimately, allow mankind to live at peace with his own race, his environment, and the natural resources so abundant and so often misused.

All of us have our own ideology. Those raised in this country believe in a certain set of principles, with some political spread to be sure, but our own historical, cultural and political background is an innate part of our thinking. We use this background as a philosophical framework within which we interpret observables as "facts." We do not stop to think too much about this process; we are too used to it. Partisans of different political parties, for example, are quick to see the opponent's framework and to label it. And we are very quick to perceive the framework of a person raised and educated in other cultures and political systems, e.g. a centrally-planned society as opposed to a market society (euphemisms for communism and free enterprise!). Thus, I beg the reader to read through Garcia's material with an awareness that there is always an interpretation present at some level, and some synthesis within an intellectual/cultural framework. If the reader is aware of this explicitly, I believe he or she will gain insights into the material which may well differ from the author's.

Extracted Findings of the Report "Nature Pleads Not Guilty" by Rolando V. Garcia

The International Federation of Institutes for Advanced Study (IFIAS) proposed in early 1972 to study in detail what was happening to the world as a result of the dramatic anomalies of climate experienced recently in several regions. The focus selected for the study, after the 1972 droughts and elsewhere, was the enormous human suffering and significant worldwide economic dislocations ostensibly triggered by these droughts.

A few months after the project began in 1976, under the direction of Argentinian meteorologist Roland V. Garcia, some new perspectives began to emerge. It appeared that the role of climate was less pervasive than had been assumed; the relationship between drought and social phenomena was far from simple cause and effect. From this point on, throughout the remainder of the 36-month investigation, the main job was to find enough evidence to confirm or deny the working hypothesis: What happens after drought has struck is much more determined by the structure of the whole ecosystem, including its social and political components, than by the drought itself.

To accomplish this task, Dr. Garcia recruited a team of physical and social scientists at work in Central and South America, Asia and Africa. Garcia clearly wanted to obtain an analysis from inside these countries where the problems occurred, one expressed by qualified people who had actually lived through this problem era. Their mandate was simple: collect as much data as possible and be sensitive, as well as cooly objective, in formulating conclusions.

The outcome of the work, compiled and integrated by Dr. Garcia, appears in more than 40 papers comprising three volumes. The first volume was published in 1980 by Pergamon Press under the title "Nature Pleads Not Guilty."

Climatic Fluctuations as Usual

It is normally assumed that during the year 1972 widespread food shortages and famines, as well as serious disruptions in the international food market, were the direct effects of extended and simultaneous droughts affecting various continents. Some speculations went even further, and one finds quite definite statements, from authoritative sources, warning about the "growing evidence that the world is entering a new climatic regime." The prolonged Sahelian drought was considered by some climatologists as indicative of such changes.

The report considered, and discarded, the assertion that we are witnessing a period of profound climatic changes. Such a contention is based on statistical analyses of mean atmospheric temperature changes. Quite apart from the difficulties inherent in the problem of climatic prediction, there are serious difficulties in reaching an agreement on the changes that have already occurred in the atmosphere in the last 30 years. A survey of the results published by several authors shows large disparities in defining the large-scale variations of atmospheric temperature. There seems to be no

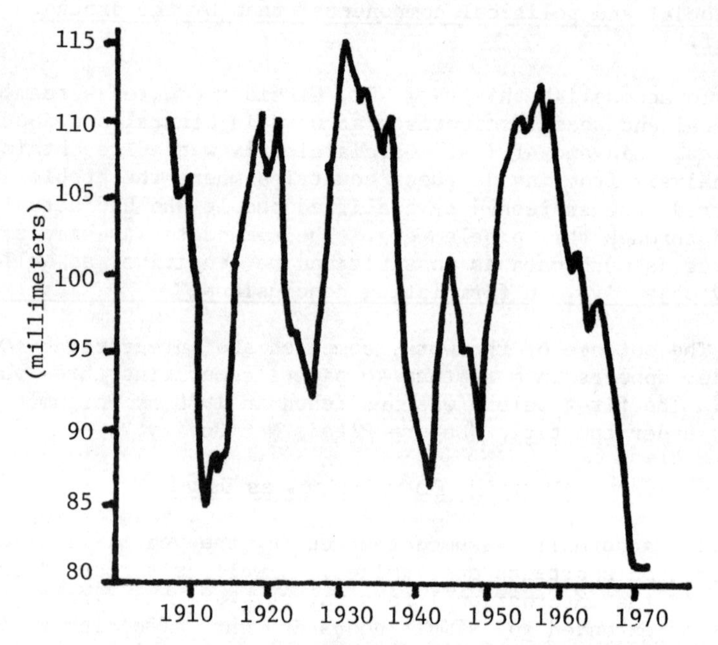

Fig. 1. Five year means of rainfall, West African marginal rainfall areas, percentages of long-term average. From Jenkinsin, A.F. "Some quasi-periodic changes in rainfall in Africa and Europe." WMO Publication No. 421, 1975, pp. 453-460.

clear evidence, then, that the amplitude of climatic variability during the recent decades is significantly different from the natural variability of the past century.

On the basis of what is known today about climate variability, any rational use of climate information must take into account three different kinds of phenomena. First, there is a normal, expected fluctuation around some mean value derived from a long period of climate history, which generally has a range we can determine from long-term records. This fluctuation has some oscillatory character, not cyclical, and is more pronounced in some regions than in others. Our understanding of this variability is still very much an open question, but we do know enough to be able to design food production systems, for example, which allow for margins of safety for downturns -- droughts in particular. Although it is not known exactly when these periods of climatic fluctuation will occur, it is known they will occur. The society should be prepared to absorb the climatic anomaly when it comes -- with some hardship, to be sure, but not catastrophe.

In Figure 1 the annual rainfall for West Africa is plotted for much of this century. A few features of this plot illustrate some of the problems of the Sahelian peoples. First, deficiencies in rainfall are not all that uncommon. The 1972 episode was the deepest deficiency in this century, but it should be noted that the rainfall amount also fell more slowly than in previous episodes. Were there sufficient advance warnings to governments starting in about 1968 to have alleviated some of the suffering? There seems to be evidence that there were responsible reports and alarms raised, and then ignored.

The drought episode in the 1940s was exacerbated by World War II. No help could be given at that time. Even though the drought was less severe than in 1972, the suffering was great, perhaps greater; there is insufficient information to make a judgement. The earlier drought in the 1910s appears, from some available information, to have been somewhat less devastating in its effects on people and herd animals in West Africa. One is lead to the interpretation that mere rainfall figures do not allow a full diagnosis of a particular drought and the ensuing societal disruption. The societal state itself must be taken into account.

One should also note the rather extended period of above-average rains just preceeding the fall to the 1972 low. This happened to be a time of emerging independence for the Sahelian countries -- shaping of new governments and national development policies, rapid expansion of commercial agriculture, need for convertible currency and development capital, and marked advances in agricultural technologies. It is said by some observers that these kinds of forces pushed the pastoralists out of lands that, owing to temporarily favorable conditions, could be exploited for short-term gains in commercial agriculture. The pastoralists, then, had to make do with lands that were even more marginal in their carrying capacity for herds. And the marginal lands that as a result of a few years of good moisture were exploited too fully, fell back to a worse state once the moisture departed. Finally, political developments took place that restricted the traditional nomadic adaptation to changes in climate -- moving to where there was forage, thus preserving herds, as well as accomplishing a natural rotation of grazing sites. All in all, while it cannot be gainsaid that a great natural deterioration in climatic conditions took place in West Africa around 1972, the resulting suffering and deaths far transcended what might have happened had not many other cultural, economic and social forces been at play.

Nothing "Normal" About Climate

One other interesting predilection of nature is illustrated in this figure. One should imagine a horizontal line through the 100 percent (i.e., normal) point on the vertical axis. Almost never in the period is the rainfall "average." It is almost always considerably above or below "normal." This seems to be quite a common feature of marginal lands. Small changes in atmospheric circulation bring or deny rain. Thus, it is not wise to calculate a carrying capacity in such lands based on "normal" climate. The climate is bimodal: 1) some rain; 2) little rain. Quite different planning approaches have to be made for such lands, approaches that take into account local political and economic factors, cultural habits of the peoples, and the resources available for using the land in different ways.

A second class of climatic variations includes the rare and large events -- great floods, very prolonged droughts, etc. We can examine historical evidence to obtain some idea of their frequency of occurrence. Needless to say, the boundary between the first and second kind of climatic variability is arbitrary, and it is a matter of judicious judgement for each socio-ecosystem to decide whether a certain range of variability is explicitly planned for. The "rare" events that are omitted from plans do, when they arrive, call for emergency measures which may include international solidarity, aid programs and regional cooperation agreements.

A third class of variation is of long trend -- a cooling

or a warming over a century or longer. In the past this type of occurrence has had strong social effects, such as migrations, or even wars to enable populations to migrate. Today we are studying mechanisms that might induce trends -- use of fossil fuels, deforestation, urbanization, concentration of energy-generating plants, etc. Our knowledge of climate-forcing mechanisms is too incomplete to give any reliable prediction at this time. But as far as the developing countries are concerned, the events in Africa, Asia and Latin America on the socioeconomic and political side are, and will continue to be, of such a magnitude that they are going to shape the future of many of those countries far more importantly than any change in climatic conditions during the coming decades. The report provides evidence of the difficulties in assessing the actual climate impact of a known nature under known social conditions, and also shows that the actual impact depends in developing countries, at least as much, and in many cases much more, on the conditions of the recipient society than on the actual magnitude or nature of the climatic perturbation.

The report's analysis, however, does not intend to minimize the importance of climatic variation, and specifically drought, for some countries. Quite to the contrary, the report emphasizes the decisive influence of a drought on the whole life of a fragile society. The release of internal structural instabilities does in fact amplify the effects of even a mild natural perturbation starting in the ecosystem. Under certain conditions, a not-very-severe drought or flood may thus have catastrophic consequences, out of proportion with the intensity of the anomaly.

Climate-Vulnerable Societies

Climatic events *per se* are not the root cause, in our times, of great disasters, famines, or increased misery. The case studies in the report provide strong evidence that droughts, long and severe as they may be, are not the sole cause or even the primary cause of internal disequilibrium in the society. They merely reveal a pre-existing disequilibrium. The evolution after the drought has struck is much more determined by the structure of the whole ecosystem than by the drought itself.

A classification of countries or regions of the world which takes into account their types of response to a drought situation ought to begin by drawing a line that separates fragile societies having high structural vulnerability from those more stable with low structural vulnerability. In the former case, drought triggers an instability which is latent in the system. The direct effects of the drought are amplified by the release of these instabilities. In the extreme cases, it may not be possible to restore the pre-drought situation, even if food or other emergency aid is brought in, once the system is taken away from its precarious "equilibrium" conditions, unless some structural changes are introduced into the society itself. A society having a socioeconomic structure of low vulnerability, on the other hand, is a stable system in which the social organization has ingrained response mechanisms to overcome the effects of either short or prolonged droughts.

Two examples illustrate the point. Example 1 is a country with an agro-system which makes it self-sufficient in food. It has, in addition, an advanced social organization and a high economic level. A drought produces a failure of the crops. But the country either has enough stocks, or is in a good position to import all the food needed. Moreover, there are social mechanisms to prevent side effects, such as unemployment or under-employment, and to provide credits for farmers and social assistance to rural workers. The country as a whole will feel some effects of the drought on the overall economy, but no direct effects such as shortage of food will be felt in any sector of the community. The social security schemes, on the one hand, and appropriate technologies on the other, will act as homeostatic mechanisms to reverse the changes produced by the drought and to restore equilibrium once the "perturbation" ceases. The system is stable. A good example was England in the 1976 drought.

Example 2 is a country with a food-producing system composed of large farms where most of the labor is undertaken by rural workers on a wage basis, and with no social security schemes. The country has food stocks available on a commercial basis. A drought or a flood partially ruins the crops. The price of food goes up and speculation begins. Inflation sets in, and salaries remain far behind price increases. The rural agricultural workers become unemployed. The food, though available, passes out of the reach of large sectors of the community. There are migrations of rural workers, to towns or to industrial centers, searching for jobs and food. The distribution of both the urban and rural population may undergo profound changes. After the "perturbation" is over, the system is not restored to the previous equilibrium conditions: once the system passes beyond the point of instability, it evolves to a new form of self-organization, i.e. a new structure.

The Sahelian countries are typical examples of evolution in recent decades towards more vulnerable conditions. The pre-colonial pastoral and agro-pastoral societies had adequate response mechanisms ingrained in the socioeconomic structures. When these began to break down during colonization and after independence, the region's vulnerability to climate increased, laying the foundation for the Sahelian catastrophe in the early '70s. China's evolution, on the other hand, provides an example of change in the opposite direction. A relatively egalitarian social order, backstopped by coordinated natural and technological resources and enhanced food production and distribution management, has considerably short-circuited massive famines caused by frequent and severe flooding or droughts.

Perceptions of the '72 Food Crisis

A climate particularly unfavorable for food production (droughts hitting several continents), a soaring demand for food spurred by continuous population growth, and rising affluence have been the reasons normally adduced to explain the food crisis that began in 1972 which resulted in declining food reserves and sky-rocketing food prices. This was the basic assumption accepted by the UN World Food Conference (1974) and a large number of well-known reports from official institutions at national and international levels, as well as influential articles and books. For the purposes of the report, the authors have termed this the "official view."

Two categories of problems which are put together in current "official view" explanations of the 1972 crisis need to be distinguished, since they belong to two entirely different realms: the events related to the profound changes which took place on the international food market in the early '70s, and the events associated with the droughts affecting several regions of the world at about the same period. Both sets of events were originally independent of each other. They interacted at a later stage, although the latter played only a minor role in the evolution of the former.

The figures on world food production and trade, curiously enough, do not support the "official view." World food production had been generally rising sharply, much faster than population growth, with two pronounced peaks, one in 1971 and another in 1973, as seen in Figure 2. In this long-term context, the relatively small drop in 1972, partly caused by a U.S. policy decision to reduce planted acreage 10 percent, cannot be deemed a "crisis," especially on a global scale. This realization led the authors to examine a broader body of information. Moreover, on the basis of the FAO and other official figures, there was adequate global food production to raise nutritional standards everywhere to accepta-

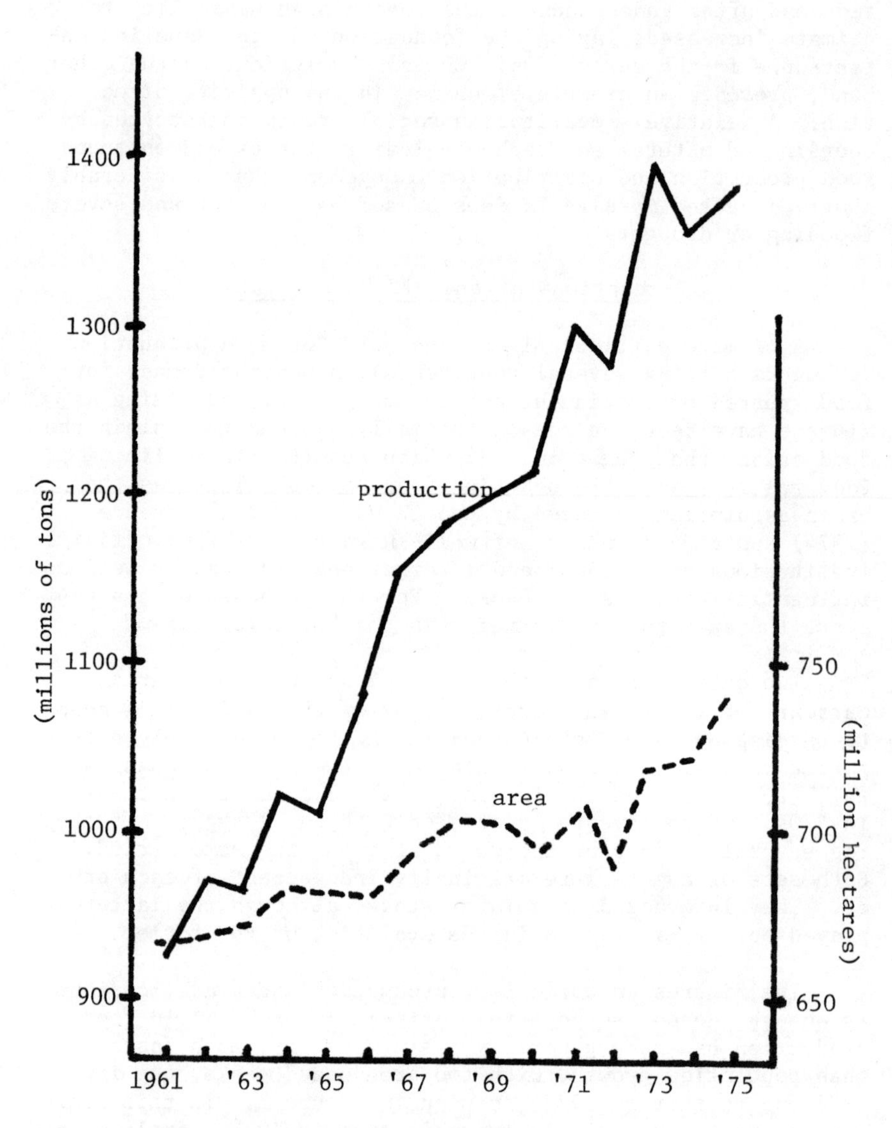

Fig. 2. World production of cereals and area harvested. From FAO Production Yearbooks.

ble medical levels. The food production potential is in advance of population tendencies, at least for this period. For the three-year period 1971-'72-'73, food production rose by about twice the amount estimated by the UN as needed to feed adequately the increased population of that period.

Food trade figures show another important factor. Many of the countries that were hard-stricken by the climatic events in the 1970s and asked for food aid, were net exporters of food. There is no other conclusion except that unavailability of food to some segments of the population was a consequence of national policies. A change in eating habits of Europeans, Soviets and North Americans, or placing fallowed land in the United States back into production, will not solve the problem of those segments of national populations who are the victims of policies, advertent or not, of their own governments and those who control the agricultural production system. The droughts only exacerbated their situation. When the rains returned, some pastures and marginal lands could again be placed into food production for the local population, who could then recuperate from starvation to be faced merely with severe malnourishment. The "productive" sectors of the national economies continued to develop and further isolate the poor sectors from participation in the various profit-making productive processes.

Further investigation and, in particular, discussions with researchers familiar with a large variety of social and political structures in Africa, Asia and Latin America, convinced the authors that availability of food to a significant segment of the population in these areas was only marginally related to world production amounts and grain trade. Indeed, for a large segment of society there is, in effect, no or insignificantly little purchasing power at the local level.

The report, then, takes the position that structures -- i.e. social organization, political systems, control of credit and capital; large-scale production arrangements such as international agro-business; international inter-governmental organizations and their bureaucracies, and the "needs" of these bureaucracies; national power sectors, their bureaucracies and their needs; cultural patterns and forces, etc. -- developed by humans into social, political and economic ways of life, have a fundamental role to play in how things work, who has access to food and other essential elements for a minimum decent life, and who does not.

Evidence has been given in several of the case studies to show that, at least in some instances, these structures only survive by paying an irreducible price: some part of society, some population segment, must remain outside the cycles of work, money flow and productivity that characterize the structure to which they belong. The production of these people plays no role in the overall figures of world food production. Their role in the local economies is, however, quite significant, and it is erroneous to think of these sectors as being out of the productive system as such. They are marginal in the sharing of benefits within the society, but well-integrated into the system as far as their contributions to the development of other sectors are concerned. Governments and all types of organizations that have the power to decide how these matters are directed, as well as economists advising them or publishing their own studies on development, may have to realize that the commonly used economic parameters are meaningless when applied to this marginalized population of the Third World.

Did Climate Escalate Food Prices?

Following the line of reasoning of the "official view," the depletion of cereal stocks and the exceptional price increase of grain and other foodstuffs were a result of the fall in world food production and the large-scale food purchases by the drought-stricken Soviet Union and a number of developing countries.

Critical analysis by the authors of the report revealed that fluctuations in the international grain trade and, in particular, the price variations, during the period 1972-'75, were the result of a changing policy concerning the structure of world food production and trade, and not the accidental effect of climatic phenomena or the inevitable consequence of a gradual increase in demand due to demographic pressure.

The key element in the alternative explanation offered by this study is the set of measures adopted in the United States by the Nixon Administration, which introduced fundamental changes in international economic policy and particularly in international trade. The history of this process is briefly reviewed on the basis of two official U.S. documents: the Williams Report (1971) and the International Economic Report of the President to Congress (March 1973). The Williams Report stressed the need for the U.S. to take advantage of its competitive position in the world market in agricultural (and high-technology) goods.

The new policy necessitated selling cereal stocks on a massive scale. Ensuant sales agreements, their effects on producers and their effects on future buyers substantially modified the rules of the game in the international grain market. The Soviet-North American grain-trade agreement was the result of an understanding from which both parties expected to derive considerable benefits. For the U.S., it meant a unique opportunity to get rid of excessive and costly reserves, to stabilize the balance of payments and to stimulate domestic production by being able to lift subsidies and restrictions, putting into practice the recommendations of the Williams Commission. In effect, then, the U.S., through devaluation of the dollar, massive sales, and other mechanisms, switched from an "aid to trade" policy.

One could still think that the 1972 drought played a major role in the sharp increase in the value of U.S. food exports between 1972 and 1975. The results of this study suggest that it did play a role, but the drought was neither the starting point of the process nor the dominant factor in the subsequent developments. In light of the analysis of this report, the sharp price rise would have occurred even in the absence of the climatic anomalies of 1972.

Factors Behind Famines and Malnutrition

Malnutrition became a central subject after it was found that there was no reliable answer to questions such as "How many people died in the Sahel during the drought?" There exists a striking parallelism between the explanatory schemes for the drought as a meteorological event, and for the famine as a social event. To "explain" a drought, one needs climatological analysis which automatically implies reference to large-scale space and time processes. It is only within this long scale that the anomaly called drought can be given a meaning and can be provided with a meaningful explanation. Likewise, famines occur as anomalies within large scale processes in societies which regulate the changing patterns at the level of nutrition. It is only with reference to this background that the famines have a clear meaning and that they can be given a significant explanation. The studies on malnutrition are, thus, the counterpart of the climatological studies.

Malnutrition is the most widespread disease in the world. The estimated population living below acceptable levels of food intake goes up to roughly 1,000 million people. Although there are rough figures on the magnitude of malnutrition in the world, no reliable information is available on the number of people dying because of malnutrition. The evaluation of the magnitude of malnutrition and the assessment of the incidence of malnutrition on mortality is an extremely difficult talk. As a general rule, in the places where there are statistics there is no malnutrition, and in the places where there is malnutrition there are no statistics. This notwithstanding, it can be shown, beyond any reasonable doubt, that the magnitude of the problem is shocking.

The vital statistics systems of the world would record, in theory, all deaths which take place in certain defined geographical areas, as well as the causes of death. In practice, however, this does not happen. There are three main reasons for this: under-registration of deaths, unavailability of health services to much of the population, and existence of biases in the current system of determining the "basic cause of death."

Examples from the case studies show that the percentage of under-registration of deaths in some Latin American countries has been as high as 40 percent and even 60 percent. The situation is much worse in Africa. In Upper Volta, the mortality rate as registered by vital statistics amounted only to one-seventh of the actual rate. In Ethiopia, figures indicating the mortality rates are not available; registration of vital rates is virtually unknown in the rural areas of the country, and even in the urban areas registration is done only on a voluntary basis.

As a consequence of the maldistribution of health services, there is a high percentage of recorded deaths whose causes cannot be ascertained, as they are not certified by physicians, and most of these deaths from unknown causes occur in those geographical areas and social classes where malnutrition is the dominant pathology. Moreover, the International Classification of Diseases by WHO has adopted a worldwide method for determination of causes of death by choosing <u>one</u> "basic cause of death" for every death. This decision tends to underestimate malnutrition as a cause of death by allocating to another disease -- usually an infectious one -- the role of basic cause of death, even if both diseases are recorded in the death certificate. Thus, an infant death "caused" by a bronchopneumonia, which is in turn caused by a measles infection in a child weakened by malnutrition, will be assigned to "measles," and malnutrition will not even be mentioned as a causal agent, even though it is well known that fatality from measles is a function of nutritional status, thus making malnutrition as much a cause of death as the measles virus.

The fact that malnutrition as a cause of death is seldom properly registered has a double consequence which cannot be overemphasized. First, the magnitude of the problem remains hidden. Second, when there is a "natural disaster" such as the Sahelian drought, the deaths produced by malnutrition become too obvious. There is a tendency to attribute all malnutrition-generated deaths, and in fact malnutrition itself, to the natural disaster. The effects of natural disasters are thus greatly exaggerated.

Moreover, there is growing conviction that increased malnutrition in developing countries is caused by population growth, responsible for the catastrophic effects of droughts or other natural phenomena on some societies. Neo-Malthusian arguments have created great alarm and led to the conclusion that there can be no significant improvement in per capita food supply in developing countries without declines in birth rates. The report maintains that although demographic pressure will be in the long run a serious problem for mankind, it cannot be held responsible for any of the national or regional catastrophes of recent times, including those of 1972. The report provides numerous examples of countries which are often mentioned as suffering from food shortages due to excess population, and yet are net exporters of food even in periods of famines.

International awareness of the semi-permanent crisis affecting large segments of the population in developing countries unfortunately does not go beyond the recognition of certain symptoms, and very seldom is the real cause explored. The analysis remains, thus, at the level of external factors. When looking for the "culprit" among them, climate, population pressure and environmental problems come to the fore. The emphasis is placed on the production/population ratio or on the misuse of the soil, and the diagnosis misses the point as the effects are taken for the cause and vice versa. Even when, penetrating into deeper waters, the emphasis is shifted from production to distribution of food, the analysis remains superficial as long as the distribution problems are taken to be those associated with transport and storage, instead of those related to the accessibility of the food. The actual problem is the distribution of means which make the food accessible to the people.

Roger R. Revelle

3. Uses of Climatic Knowledge in the Food Systems of Developing Countries

Introduction

In developing nations, climate has its greatest impact in the zones of widest climatic variability -- the warm, semiarid and subhumid zones, particularly the regions where monsoon rains prevail. These include nearly all of the Indian subcontinent; the semiarid and subhumid regions of China; Thailand and parts of Indo-China; much of the Middle East; part of Africa north of the Sahara; the Sahelian zone south of the Sahara, extending from Senegal through the Sudan to Ethiopia; parts of southern Africa; northeast and southwest Brazil and parts of the central Brazilian plateau; northern Mexico; large areas of Argentina and Chile, and the coastal zone of Peru. Present populations in these regions total more than 1500 million people -- more than 50 percent of the developing world.

With the rapid growth of populations in developing nations, a sustained high level of agricultural production is becoming more and more critical. All resources are being stretched closer to their limits, and climatic variability is an increasing threat to the survival of people at the margin of subsistence. Fortunately, the climates in most of these regions are warm, so that crops can be grown throughout the year, provided sufficient water can be made available.

Three kinds of knowledge about climate are important in the less-developed countries: 1) the statistics of climatic variability, determined by suitable measurements over time and space, combined with the analytical manipulation of these measurements; 2) the ability to make short-range probabilistic forecasts of what the weather is likely to be a month to a year from now, and 3) the probability of climatic change -- i.e. the probability at some time in the future of a change in the mean state of the climate and/or in the frequency distribution of different states. From the standpoint of decision-makers, the first kind is the most important.

How Farmers Assess Climate Impact

Decision-making in agriculture occurs primarily at four levels: the individual farmer, the agricultural sector in a particular country, the national level, and the global society as a whole.

In traditional peasant agriculture, the farmer's decisions as to what crops to plant and when to plant them were made long ago, over many generations, by a combination of trial and error and canny observation of local climatic variability. The principal basis of these decisions was risk-aversion -- the necessity to produce enough food for survival in years when the climate was "bad." Market forces had little influence because the crops were not sold, but consumed within the village. And few decisions were necessary at the sectoral, national, or international levels.

With the unprecedented growth of populations in developing countries during this century, a transformation to market agriculture has become necessary. This means that the farmer must buy fertilizers, farm tools and machinery, and other inputs in order to increase his yields per acre, and he must sell a large part of his crop to pay for these inputs. With market agriculture there is much more flexibility, and the farmer must make decisions each year as to what crops to plant, when to plant them, the quantity and kind of inputs to purchase, and when to sell his harvest. Moreover, decisions need to be made at sectoral, national, and international levels, to ensure as adequate a food supply as possible for large and growing populations.

The farmer's decisions will be better if he can take account of the probability distribution of the timing and quantity of water supplies, potential evapotranspiration, and solar insolation. In a region of low and uncertain water supplies, he will plant millets instead of maize in the African Sahel and groundnuts instead of rice in Indian Saurashtra. He will be reluctant to borrow money to buy fertilizers. Where the danger of floods during the monsoon season is high, the Bangladesh farmer will plant floating rice, or nothing at all, and concentrate his efforts in the sunnier, more dependable winter seasons. If even moderate rainfall is likely in April and May, the Pakistani farmer will not plant cotton, because the cotton will waste its photosynthetic production in vegetative growth, and what bolls do develop will be severely damaged by pests stimulated by the moist environment. In regions where the monsoon starts and stops intermittently, the farmer will hold back half or more of his seed, to be able to make a second or third planting if the first planting fails for lack of moisture.

Agroclimatic information collected and interpreted by competent interdisciplinary scientists will be especially useful at the farm level in areas of rainfed agriculture, where crops that are not traditional for a particular region, such as fruits, nuts, vegetables, soya beans, and oil seeds, are being introduced for the first time. At the sectoral level, the agroclimatologist can give climatic specifications to the plant breeder who is attempting to produce better-adapted varieties.

Climate-Responsive Water Management

In most of Asia, rainfall and runoff are concentrated in the few months of the monsoon season. Irrigation and drainage development and flood protection, at a total estimated cost of more than $100 billion, are essential if agricultural production is to be increased sufficiently to feed the growing population during the next 25 years. Besides providing water to farmers in the dry season, thus permitting two or three crops to be grown during the year, surface reservoirs and wells will smooth out the variability in water supplies within the rainy season. If the underground aquifer is sufficiently large, it will also be possible to reduce year-to-year variability by pumping down the water table during the dry years and recharging the groundwater during wet years.

Accurate information on the frequency distribution in time and space of rainfall and runoff throughout a river basin is obviously essential for the design of irrigation, drainage and flood-control systems. But it is less obvious that such information is also necessary to optimize the management of these systems after they are constructed. In operating a surface reservoir, for example, the principal action involving a human decision is to release water through the penstocks and electricity-generating turbines or the irrigation tunnels in the dam. Though the action is simple, the decisions as to when and how much water to release throughout the year are complex. They will depend on often conflicting demands for irrigation and for electric power, on the economic and political weights assigned to these two uses, and on the best available estimates of river runoff into the reservoir during the remaining months of the year. All three of these decision factors rest in part on climatic statistics.

In the management of an underground reservoir, the mean depth to the water table should be maintained at an optimum level, such that water-logging and high evaporation rates will not occur during one or several years of heavy rainfall or river runoff, while pumping costs will be kept as low as possible after one or several dry years. Again, information on year-to-year climatic variability is essential.

At the national and international levels, the size and location of food-storage facilities, the quantity and kind of food reserves, the tonnage of ships for international food transfer, national pricing and procurement policies for agricultural products, and planning and institutional development for imports and exports should all be (but seldom are) based in part on both regional and global statistics of climatic variability.

Would Climate Forecasts Help?

Would climatic forecasts over one to 12 months help decision-makers at different levels? Under present conditions, such forecasts might be least useful to farmers and most useful at national levels. What does the farmer do with a forecast that says there is 75 percent probability that next season's rainfall will exceed 400 mm, instead of the 50 percent probability given by the statistics? Doubtless he should do something, e.g. plant a more water-sensitive but potentially more profitable crop or increase the amount of fertilizer that he will apply. But he is not likely to do either of these things unless sectoral or national actions are taken to assure the prices he will receive if he has an abundant harvest and to insure him against crop failure if the forecast turns out to be wrong.

At the sectoral level, climatic forecasts would enable fine tuning of the decision rules for release of reservoir waters. Forecasts of higher than normal rainfall or runoff should also stimulate planning and mobilization of measures for flood control. Forecasts of future weather conditions leading to increased pest populations would be helpful in organizing plant-protection measures.

At the national level, both regional and global agro-climatic forecasts would be useful in setting procurement prices, planning and allocating resources for food imports, or planning and organizing for exports of agricultural products. At the international level, no institutional mechanisms for using global forecasts and little realization of the need for them now exist. But this does not mean that such institutions are not needed.

There is, however, a distressing shortage of data concerning hydrologic and other climatic parameters in most developing countries. Often only a few limited types of observations of questionable accuracy are available for only a few years. One of the important problems of climatology is to find and apply methods of data analysis that will make it possible to use short series of fragmentary, innacurate measurements in helping to improve the food systems of developing countries.

Harnessing Climate for Energy

Planners and decision-makers in the less developed nations should consider use of climatic information for the development of energy sources for agriculture. Maps of the monthly distribution of wind velocities, for example, would define areas where wind energy can be effectively and economically deployed -- e.g. for pumping irrigation water. The number of days per month in which the average wind velocity is less than two or three meters per second (below which windmills characteristically do not work) would determine whether wind power might be a practical energy source. The balance between thermal and hydroelectric power plants in any region depends on the frequency and length of droughts that reduce the water supply to reservoirs and increase the demands for power for pumping irrigation water. The 1972 drop in agricultural production in India was in part due to the fact that the nearly empty reservoirs could not supply sufficient electric power for pumping, and there was too little backup from thermal power stations.

Development of biomass energy for use in producing and processing food, either from agricultural residues or from fast-growing trees, must take account of both average climate and its variability. For example, the anaerobic bacteria that convert animal and plant wastes into methane in biogas converters operate efficiently only over a narrow temperature range. Biogas converters will not be very useful when the annual temperature range is too great. Selection of fast-growing tree species for energy plantations depends on the balance between water supply and evapotranspiration at different seasons of the year, on humidity, annual temperature regime and soil-water relationships, and on the variability of all these factors from year to year.

Climate and Fish Production

Several fisheries in less-developed countries -- notably the anchovy fishery off Peru, the shelf fishery off India's Malabar coast, and Korea's wide-ranging distant-water fishery -- appear to be more or less critically dependent on variation in oceanographic conditions, i.e. in the ocean climate. For example, Peru's anchovy catch went from 12 million to less than 2 million tons in one year following the occurrence of the oceanographic phenomenon called "El Nino," an occasional invasion of warm water into the usually cool waters of the eastern equatorial Pacific. While most fish production in less-developed countries comes from shallow-water nearshore or freshwater fisheries, the latter often in man-made ponds, the fact that fish provide about 10 percent of the protein available to the present world population (and a much higher percentage in many poor countries) underscores the necessity for further study of climate's effect on fish productivity.

Finally, Man's Impact on Climate Itself

At present, our level of understanding of possible long-term changes in the main state or the year-to-year variability of the global climate is low. For example, we are unable to estimate changes in patterns of precipitation and runoff with an increase in average air temperatures which might be brought about by an increase in atmospheric carbon dioxide. Under these circumstances, planners and decision-makers in less-developed countries cannot be expected to take much account of these possible changes. The world community, however, should act to ensure that sufficient planning and investments in irrigation, drainage, and energy conversion are undertaken to prevent a deterioration of per capita food supplies that might result from adverse long-term climatic changes brought about by the profligate use of energy in developed countries.

The leaders of less-developed countries can be expected to take greater responsibility for those actions by their own citizens that may be causing deterioration in the regional hydrologic cycle. For example, destruction of forests in the Himalayan foothills of Nepal, India and Pakistan is probably resulting in more frequent and more severe floods in the Indo-Ganghetic Plains, on which more than 400 million people depend for their food supply. Equally serious, although less certain, may be a steepening of the river hydrographs, i.e. an increasing concentration of annual river flows during the four months of the monsoon season, because of destruction of water-holding vegetation and soils. This may diminish surface water supplies for the winter crops.

Martin E. Abel, John A. Schnittker,
Diane C. Brown

4. Climate-Defensive Policies to Assure Food Supplies

Introduction

Transfers of food among countries through commercial trade and food aid have been the major vehicle for stabilizing food supplies in nations affected by unfavorable weather and climate, and by national policies. Yet serious problems hinder the effective transfer of food, both for meeting normal domestic food needs and in emergency situations. The ability to identify food needs and the responsiveness of national and international institutions suffer from a lack of adequate and timely information. Financial, educational and institutional constraints often thwart stabilization of food supplies, and national food distribution systems are often inadequate to provide needed food. Both a national food reserves policy and the structuring of bilateral trade agreements offer effective methods for ameliorating food shortfall situations and stabilizing food prices.

Sources of Instability in Grain Production and Prices

Climate-Related Instability

There is growing evidence that weather and climate fluctuations have increased in frequency and severity during this decade, resulting in greater instability in national and world food supplies. Developing countries in particular have been vulnerable to this instability because they lack the financial, institutional and technological resources to mitigate the impact of reduced food supplies and higher food prices. Fluctuating food supplies are especially threatening to developing countries, because high rates of population growth are outpacing rates of growth in food production.

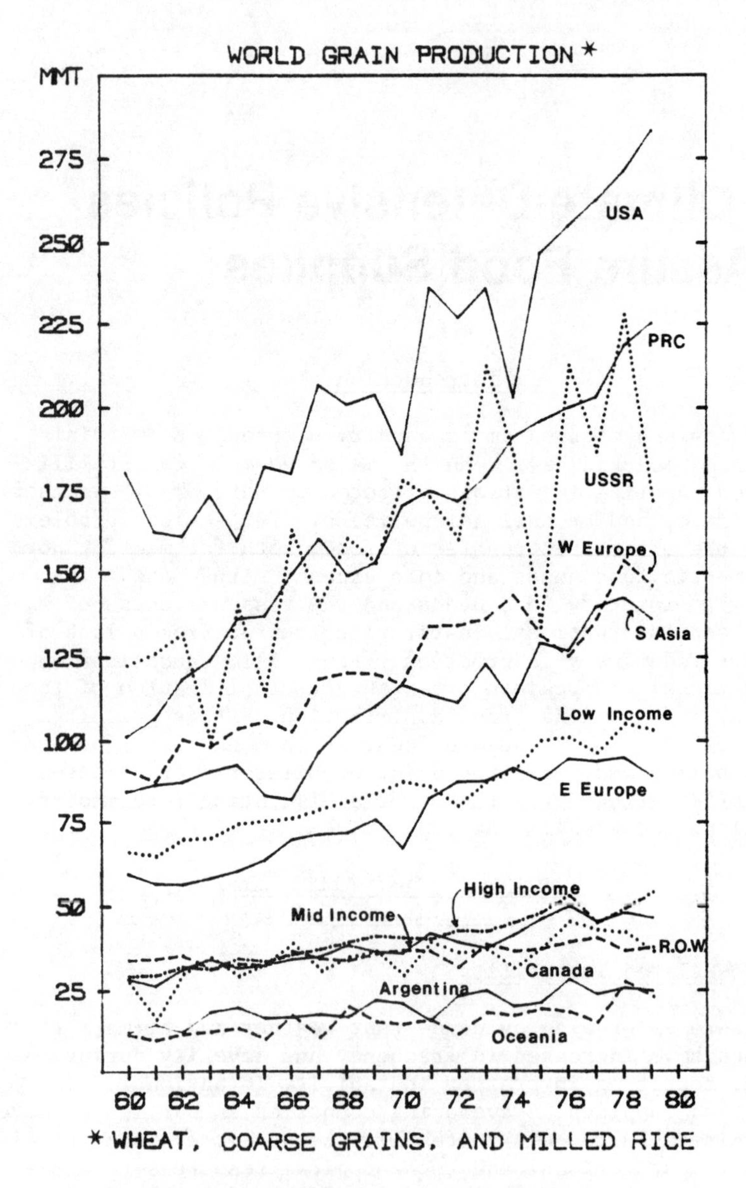

Fig. 1. World grain production 1960-80. Source: Schnittker Associates (based on FAO data).

This leaves many segments of the population near subsistence levels, with very little cushion against malnutrition, and even starvation, in times of food shortages.

Concern about the effects of climate variability has increased dramatically since 1972, because of a series of weather-related events that had serious effects on world food production. In 1972, crop failures in the USSR and India were major factors in the one percent decline in world crop production that resulted in sharply-reduced food reserves. The continuing of the Sahel drought into 1973, a poor 1973 U.S. corn crop, production declines in the USSR in 1974 and 1975, Asian monsoon failures, droughts continuing into 1976 and 1977 in several important producing regions, and the 1979 Indian grain shortfall all emphasized the degree to which major crop-growing regions are affected by weather conditions. This is shown graphically in Fig. 1.

A significant aspect of the post World War II food situation is the extent to which the global food supply depends upon the giant producing countries or regions of the U.S., Western Europe, the USSR, China and South Asia. Thus, climate-induced crop losses in these areas have grave implications for developing countries dependent upon imported food.

Finally, while climate-induced production shortfalls in major grain-producing and exporting countries can seriously destabilize world supplies and prices, weather-related local crop disasters can be especially devastating to the populations of developing countries, especially for those rural, low-income peoples who suffer first and foremost from localized crop failures.

Destabilizing Effects of National Policies

Countries have often taken actions to stabilize their domestic food price levels that have the effect of destabilizing world markets. If, during a period of short supply in the world, a nation limits exports of a commodity in an effort to stabilize prices at home, prices on international markets will rise. If an exporting nation supports the price of its product in the domestic market in a period of general surpluses, and dumps the excess indiscriminately on world markets, prices in international markets will decline. Such flagrant actions are so widely condemned that nations rarely take them. World opinion and the expectation of retaliation are effective deterrents. In an extreme emergency, however, any nation will do what it thinks it must. U.S. export embargoes on grains and oilseeds in the 1970s are a case in point.

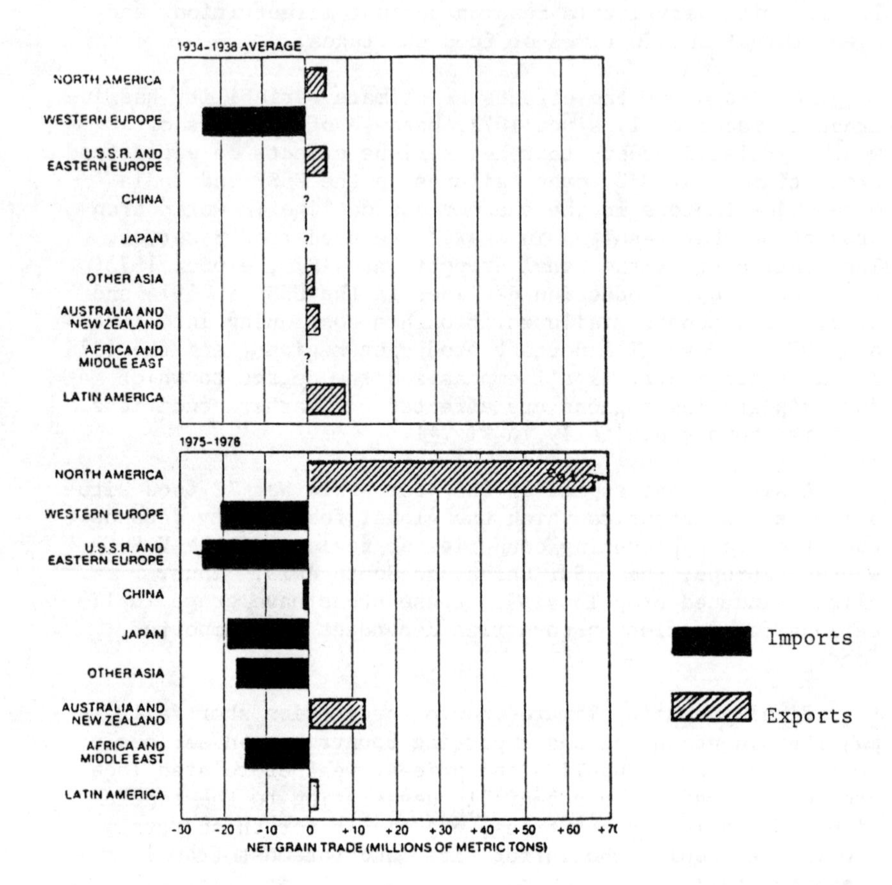

Fig. 2. The World's increasing dependence on grain exports of a few countries is shown by this comparison of the trade pattern before World War II with the situation most recently. Before the War, most regions exported grain; Western Europe imported it. Now, the U.S. and Canada supply most of the grain to make up deficits. Source: Falcon, Walyer, P. "Food Self Sufficiency: Lessons from Asia," in International Food Policy Issues, A Proceedings. U.S. Department of Agriculture, Economics, Statistics and Cooperative Service, Foreign Agricultural Economic Report No. 143, Jan. 1978.

Another case of unilateral country action is less agressive, but is becoming more common. In this situation a country insulates itself and its economy from the international market by various institutional means. Member nations of the European Economic Community and all state-trading nations, especially the USSR and China, fall into this category, as do some other countries. In a situation of world shortage and rising prices on the world market, no price signals are given to consumers in the insulated country to use less or to producers to produce more. The full burden of adjusting to the shortage falls on those importers and exporters who are integrated into the world market. A comparable situation occurs when there is a surplus: no signals in the form of lower prices are given to consumers to use more of the cheapened product.

The adjustment impact on those countries integrated into the world market would be lessened if more countries lowered their trade barriers or allowed the price mechanism to ration supplies. Certainly these approaches should be pursued, but there are definite limitations. Some economic systems preclude regular use of the market. Several market economies have effectively insulated their food and agriculture prices from world markets. Countries integrated into the world market must find ways to offset the insularity of other countries, while at the same time working towards making some national markets more responsive to world price fluctuations.

World Trade and Aid

Between 1960 and 1977, world grain trade increased substantially, growing at an annual rate of 3.4 percent from 1960 to 1971, and by 5.4 percent per year from 1971 to 1977. While the considerable trade in coarse grains reflects the growing demand for livestock products, overall trade figures reveal decisions by many countries, notably the USSR, Eastern European countries, and China, to offset climatic variability and to stabilize domestic food supplies via imports (Fig. 2).

In the developing countries, grain imports have increased fourfold since the early 1960's, but the share of food aid in these imports, primarily from the U.S., has declined. A method of food transfer for nations unequipped to purchase on the world market, food aid fulfills three basic requirements: it helps countries meet shortfalls in domestic production (emergency or disaster aid famine relief); it offsets a temporary balance of payments problem, and it supports nutrition intervention and economic development assistance programs.

Beginning with the 1966 Food for Peace Act, U.S. food aid emphasis shifted from disposing of surplus commodities to feeding hungry and malnourished people, encouraging economic and agricultural development, and building commercial markets abroad for U.S. products. These objectives, plus the worldwide build-up of surplus stocks in the 1960s, gave new impetus to efforts to increase food aid. By 1969, food aid shipments from all donors had reached 12.8 million metric tons (mmt). However, declining grain stocks after 1972 reduced U.S. food aid under P.L. 480 to the lowest levels ever in 1973 and 1974 because of domestic supply considerations and high prices. By 1973, food aid shipments from all donors had fallen to 5.9 mmt.

Current world food aid commitments, about 25 percent of estimated total imports of developing countries, now total 9.8 mmt per year, just below the minimum global target of 10 mmt set by the World Food Conference in 1974. The International Development and Food Assistance Act of 1974, the World Food Conference, and the Food Assistance Act of 1977 all indicate that the U.S. and other developed countries now recognize that production shortfalls and high grain prices are most injurious to the world's poorest people, and that past levels of food aid have not usually been sufficient to protect these people from starvation and malnutrition.

When agricultural commodities are scarce and prices are high, the developed countries have little incentive to provide high-valued commodities on concessional or grant terms. But the droughts in India in 1965-'67 and 1972, the persistent Sahelian drought of recent years, and wars in Nigeria and Bangladesh all created critical food needs that were met without regard to world supplies or prices. Sometimes, however, food needs are not met in spite of humanitarian commitments and for reasons other than price and availability.

Information and Decision-Making Problems

There are several reasons why these commitments are not met. First, there are serious gaps in information about world crop production and stocks and the potential for food deficits. Strengthening and improving national and international food information systems will not increase the supply of food in a year of crop shortfalls, but more timely and reliable crop information, and more understandable descriptions of agricultural policies would permit adjustment processes in both importing and exporting nations to begin earlier and to be pursued with great vigor. Some developing countries have made great strides in improving the ability to identify their food needs by obtaining timely estimates of crop production, improving estimates of needs or demands nationally and by region, and developing the capacity to integrate information about domestic supplies, demands, and import requirements. The improvement has come from more current and accurate government statistical programs and closer observation of crop conditions by government officials and private interests. But too many developing countries still have little timely information about domestic food conditions and what it means for food imports.

Second, even countries where domestic supply conditions are known are often unable to act on import requirements on a prompt basis because bureaucratic and political factors often slow the decision-making process needed for responding to known food needs. For example, the governments of the Sahelian countries did not make an appeal for food assistance to the international community until the spring of 1973. By that time, drought had reduced food production in these countries drastically for five years. The bulk of grain imported in response to this appeal did not begin to arrive in the drought-stricken countries until 1974.

The relief effort, with more than 20 countries providing some $150 million in aid (one-third from the U.S.), revealed serious flaws in the organization of international relief to these countries. The U.S. Agency for International Development (AID) and the UN Food and Agriculture Organization (FAO) were the principal sources of aid during 1973, but they were often unable or unprepared to act quickly and at a level sufficient to forestall further tragedy. The principal failures of this relief effort were the inattention to early warnings, the absence of advance planning, and the lack of systems to monitor and coordinate relief efforts. While there was ample warning of the Sahelian drought, bureaucratic delays and red tape, institutional inertia, and rivalries among agencies and officials marred the relief effort and kept it from being as effective as it might have been.

International political factors also have limited responsiveness to food needs. Until the early 1970s, India was a large grain importer; however, when U.S.-India relations deteriorated after 1971, U.S. food aid was sharply reduced.

If food transfers -- trade and aid -- are also to be used to reduce malnutrition in developing countries, they must be reliable and flexible enough to supply the quantities and types of food needed. It also means that greater emphasis should be placed on using food aid to promote economic development and improved nutrition. However, the willingness of developed producing countries to respond on a continuing basis has thus far been related to supplies of grain, and food aid has been highly erratic. While many countries now seem committed to meet the aid allocation of 10 mmt set by the 1974 World Food Conference, it remains to be seen if lower grain stocks at some future time will undermine this commitment.

Further Constraints to Food Transfers

Financial

Even if developing countries had the capacity to identify external food needs and if food were available, other constraints in many countries thwart stabilization of food supplies. Almost all developing countries face severe financial constraints most or all of the time. Any decision to spend unexpectedly large amounts of foreign exchange on food imports is a difficult one. Typically, developing countries have limited and unstable foreign exchange reserves. Their export capacity is limited, and developed countries' trade policies often discriminate against what they have to sell. Moreover, servicing foreign debts places a large claim on foreign exchange reserves. In many developing countries, debt servicing nearly equals the amount of new foreign aid. And finally, almost all developing countries have development plans that require capital imports, and most projects with large import components take many years to complete. If these imports are reduced to meet food import needs, projects are delayed and this is costly both financially and in terms of foregone future output.

Educational and Institutional

The typical developing country does not have the trained personnel or the institutional mechanisms required to integrate food imports with domestic production in order to stabilize food supplies. Most do not have the means to procure and store grain at harvest, to be used when food supplies are limited. Many developing countries also do not have policies and mechanisms for distributing food to those people most affected by food shortages. It is easiest to reach the urban consumer, but the people usually most affected by food shortages are the rural poor, particularly in areas where crops have failed.

Waste and discrimination in the distribution of food have often captured the attention of the press. Reports indicate, for instance, that ethnic and political rivalries in some African countries led to major inequalities in the provision of food to the nomadic peoples suffering most from the Sahelian drought.

Another real problem is the lack of basic information and research needed to effectively manage domestic and imported food supplies in developing nations. Until this information is generated, it is difficult to improve food supply management. The following questions need to be addressed:

1. What is the optimum distribution of food stocks within a country to ensure adequate distribution in time of need?
2. What are the alternative levels of food supplies and alternative government food distribution policies?
3. What are the seasonal food needs of different areas and groups within a country and are these being met?
4. What is the relationship between variations in food supplies and food prices?
5. What level of reserve stocks would be required to keep price variation within specific limits?
6. What are the likely import needs associated with different levels of domestic production? How do these levels relate to alternative levels of local stocks?
7. What are the storage losses of food under different types of storage? To what extent can these losses be reduced?

Food Distribution

In most developing countries, the food distribution systems best serve large urban areas, whether from domestic production or imports. Since urban populations are better organized politically, they are able to protest against food shortages or high food prices, and therefore receive a disproportionate amount of attention. Urban areas also have better storage and distribution facilities, making it relatively easy to manage food supplies.

A corollary is that rural areas, with 60 to 80 percent of the population and most of the poor people, are less well served. Fluctuations in production due to weather and climate variations bear directly on rural food supplies and are usually not offset by imports. Any food assistance programs for the poor rarely reach the rural population. The problem for rural areas is compounded by the fact that food supplies and purchasing power are highly seasonal, related to the crop production cycle. Most of the rural poor are landless laborers or small farmers with insufficient land and capital resources to earn an adequate living from farming. For them the food situation may be precarious over many months in years of normal harvest. It can be disastrous in years of poor crops.

Developing nations can do several things to improve food distribution in both rural and urban areas. When food reserves are established, they can be located at strategic points to ensure that all needy segments of the population have access to them. In addition, food assistance programs for the poor could be extended from urban to rural areas. In many situations, food can be provided through programs that generate additional employment and income. There is ample opportunity in many rural areas to operate public works programs to construct the infrastructure needed for improving food production, storage, and distribution. Many of these activities are labor intensive, requiring relatively little capital. They include building or improving irrigation systems, rural transportation facilities, and grain storage facilities, and reforestation projects.

Finally, there is ample opportunity in many developing countries to reduce pre- and post-harvest food losses, thereby increasing the effective food supply. While many estimates of the extent of food losses are highly exaggerated, losses are nonetheless significant. Better control of insects, birds, and rodents in the fields and improved grain storage and processing could reduce these losses appreciably.

All of the above measures would require improvements in the food policy decision-making process because specific food distribution strategies are involved and would have to become part of government policy. Countries will also have to develop the capacity to manage food supplies in order to get food to those who need it the most. This will require not only the physical capacity to store and distribute food, but also the trained manpower to implement food policies.

Grain Reserves

Grain reserves appear to be the most effective means to date for stabilizing food supplies in the short run, yet few countries have had explicit reserve policies, and progress toward this end at an international level has been limited.

There are compelling reasons for both developed and developing countries to carry grain reserves. Claims on available food when supplies are short are a crucial part of the total process of allocating scarce resources. Food reserves will mean that conflicts for available food resources will be less intense. It is important that major grain-producing or consuming countries develop and operate grain reserve programs to meet the unique needs of that country. Progress to establish national reserves will provide an impetus for any international cooperation.

Financial and economic conditions, social goals, and reserve needs will differ to some degree from one country to another, so the appropriate grain reserve program undertaken by each country should reflect this difference. A grain reserve program for the U.S. or Canada, for example, would be concerned primarily with assuring foreign buyers a continuous and regular supply of grain, while at the same time maintaining a degree of domestic price stability. A grain reserve program in India would be primarily concerned with avoiding starvation and high food prices during bad crop years, although economic, monetary and political objectives would also be important. In the USSR, the primary goal might be to stabilize the size and productivity of livestock herds and to maintain domestic social and political stability. A grain reserve program in Japan might attempt to ensure the stabilization of food prices to the large urban industrial work force in the event importable supplies are limited.

Effective reserves require explicit policies for their accumulation and management to ensure that in times of good crops some grains will be held off the market to be used when crops are bad. The willingness of a country to do this is the essence of a grain reserve policy. Furthermore, grain reserves are part of a larger set of food and agricultural policies that may involve supporting producer prices to ensure adequate incentives to expand output, and providing food on subsidized terms to the poorest segments of the population.

India: An Exception

Only India among the developing countries has created and implemented a policy of national grain reserves. India's approach could serve as a model for other developing countries, especially since it integrates policies to support producer prices with policies to stabilize prices and supplies to consumers. Annual variations in domestic consumption of food grains in India have been smaller than changes in production. Grain imports and domestic stocks have contributed to the balancing of consumption and production. India's grain reserve is especially important in 1979-'80 because grain production fell sharply as a result of inadequate rainfall.

Until 1976-'77, India was a significant importer of

grains. The annual level of imports has varied considerably since 1960-'61, reaching the 9-10 mmt level during the two bad drought years of 1965-'66 and 1966-'67, and falling to negligible levels in recent years. India was heavily dependent on food aid until the early 1970s, with most of it coming from the U.S. After 1971-'72, India moved away from aid as a source of imports to commercial purchases. Good grain crops in India in the 1975-'78 period have increased stocks to around 20 mmt, and have permitted India to become a small net grain exporter during the two recent years. The accumulation of grain stocks was not the chance result of a run of relatively good weather. It is related to the evolution of a grains policy that started about 1965, and culminated in an explicit grain reserve strategy in more recent years.

Starting in about 1965, India adopted a policy of supporting producer grain prices in order to provide production incentives. It established the Food Corporation of India (FCI) to implement its price support programs to store grain and to distribute it through domestic food assistance programs. The FCI stands ready to buy grain from farmers at announced prices. In years of large crops, prices are prevented from falling to undesirably low levels and the FCI accumulates grain stocks which then become available in years of low production, such as in 1979-'80.

India's current grain position will permit it to withstand significant declines in domestic production without reducing consumption levels. The large grain stocks will also cushion the need for imports and provide India with flexibility in timing purchases on the world market.

Bilateral Trade Agreements

In a world where multilateral arrangements are of limited (token) size and scope, much more can be done bilaterally to harmonize the interests of grain-exporting developed countries and grain-importing developing countries concerned with stabilizing external grain supplies and prices.

Long-term bilateral agreements or contracts for staple commodities have the advantage of providing both the selling and buying nations certainty with regard to the quantity of the product involved for the period of the agreement. By their very nature, long-term bilateral agreements provide assured markets and sources of supply at stable prices for the countries involved, as long as they are honored. But if one party to the contractual arrangement guesses wrong about future market developments, a long-term contract can be an embarrassment. Cancellation of certain arrangements in the sugar trade several years ago provides an excellent example of the possible results of such a situation.

In an unstable world, many nations are reluctant to enter into bilateral arrangements of several years' duration. However, when such arrangements represent "best efforts" agreements, the possibility of cancellations or defaults is minimized. Such arrangements seldom represent binding long-term contracts that fix quantities and prices. Canada, Australia, and Argentina have had certain arrangements with the USSR and China for many years. Since 1975, the U.S. has had a bilateral agreement, but not a firm sales contract, to provide grain to the USSR. Also, Thailand and Japan have had bilateral arrangements for corn for many years.

How Bilateral Agreements Can Work

For practical purposes, there are three grain-exporting countries that represent the major source of external food supplies to the developing countries. These are the U.S., Canada and Australia, with the U.S. being the dominant supplier. The U.S. is used here as an example, recognizing that the same concepts are applicable to any other grain-exporting country.

The U.S. could develop long-term bilateral arrangements with developing countries to ensure them an adequate supply of food imports. These agreements could be structured so that each importing country agreed to take a minimum amount of grain each year. At the same time, the U.S. would guarantee that country up to a maximum amount of grain in any one year. This feature of an arrangement would be similar to the U.S.-Soviet Union one currently in effect. The minimum and maximum amounts of grain would be related to historical and prospective import needs and to the grain reserve strategy of the importing country. The minimum level of imports could be used to increase grain stocks in years of good production in the developing country, thereby reducing the level of import needs in subsequent years of poor production.

Bilateral arrangements of this type would have beneficial effects for both parties. The assurance of markets at least equal to minimum import needs would provide the U.S. with incentives to hold larger grain stocks. These would be justified on the basis of commitments to meet the grain import needs of the participating developing countries. The developing countries would be assured a supply of grain in time of need. Further, there would be incentives to use imports to build and manage grain reserves in the importing countries.

The bilateral approach can also be tailored to deal with the price instability issue and the ability of the recipient country to pay for imported grain. This could be done through integrating U.S. food aid and export credit provisions, as outlined in the P.S. 480 program into the bilateral arrangements.

Provisions in U.S. Program

The U.S. P.L. 480 program contains three provisions. Under Title I, commodities are sold to developing countries on liberal credit terms. Poor countries receive 40-year loans to finance Title I P.L. 480 sales, with no down payment required, a 10-year grace period on loan repayment with a two percent interest charge during the grace period, and a three percent interest charge thereafter. For more prosperous developing countries, the loans are for 20 years, require a 20 percent down payment, and have a grace period of only two years. The interest rates are the same as for poor countries. While Title I sales are not free to the recipient country, the actual cost of food is much less than if the country had to purchase it commercially.

Title II provides for outright grants of food to developing countries. Title III permits food to be used for development purposes. It is in effect a grant program, because loans under this title do not have to be repaid.

Grains would be purchased from the U.S. under the bilateral agreement on a competitive basis. The importing country would be free to accept the best offers, so that tying purchases to the U.S. would not place the importing country at a competitive disadvantage. But not all purchases would have to be on commercial terms, even though the grain would be purchased competitively on the open market. The U.S. could negotiate a mix of commercial and concessional sales appropriate for each country's level of development and ability to pay. For example, in the case of a very poor country, the basic package might consist of 20 percent commercial purchases, 25 percent donations, and 45 percent P.L. 480 sales on long-term credit. The mix of commercial and concessional sales would vary among countries.

This "mixing" approach means that certain safeguards against sharp and unexpected price increases could be built into bilateral arrangements with developing countries. Thus, when world grain prices rose above a certain level, greater proportions of grain sales to participating developing countries would be on concessional terms. The extent of concessional sales in the total mix would increase proportionately with price rises above a certain level.

Each grain-exporting country could pursue this approach with respect to developing countries. In addition, exporters might wish to compete among themselves relative to the terms they offered the developing countries. The trade-off for the exporters would be assured markets vs. concessions to needy countries. The exporters who offered the most attractive terms would presumably be favored by developing countries.

If only a few countries were involved in bilateral arrangements, the inducement to hold additional stocks would be small and would not contribute much, in itself, to stabilizing world grain prices. However, as the number of countries involved in such arrangements increased, the supplying countries would be encouraged to carry larger grain stocks in order to honor export commitments, and this would result in greater price stability.

The bilateral approach should not be viewed as a complete substitute for international efforts such as the World Food Program and a system of nationally-held reserves. However, it is an appealing way to move toward meeting food needs of developing countries with a minimum amount of international negotiations, political road blocks, and bureaucratic delays.

Concluding Remarks

Regardless of the success of long-term efforts to increase world food supplies, climate fluctuations, changes in food and agricultural policies, and economic factors will cause short-term instability in these supplies. Such instability will have to be reckoned with in order to obtain a degree of world food security. Of immediate concern are the many food-deficient nations or regions where climate instability can and does cause severe food shortages and imposes severe hardship on millions of people. Even though the world food situation is more comfortable today than it was in the 1972-'76 period, there is no cause for complacency. World grain stocks per capita are smaller today than they were in 1972. And the mechanisms for transferring grain from nations that have it to those that need it on an emergency basis is little, if at all, better than it has been in the past. In sum, we may not be that much better equipped to deal with large declines in grain production in 1979 or 1980 than we were in 1972, and we are probably not yet in a position to protect developing countries against two or more years of poor world grain crops in a row.

There are things countries can do individually and collectively to improve the situation. One is to build grain reserves. It is in the interest of most developing countries to maintain some level of grain reserves. Even though the level of reserves that developing countries can afford to carry will likely fall short of stabilizing their food supplies, these reserves nonetheless will help cushion the effect of poor crops. Individual countries will also have the flexibility of adjusting the composition of their grain or food reserves to suit the preferences of their consumers.

Further action should also be taken under international auspices to stimulate the building of grain reserves by both developed and developing nations and to adopt mechanisms and rules for coordinating their use on a global scale. However, little progress will be made on the international front until there is a greater recognition of the importance of reserves at the national level. Progress at the national level is more likely to stimulate international action than the other way around.

There is considerable merit in bilateral agreements between grain-exporting countries and grain-importing developing countries that encompass a mix of commercial trade and food aid. The terms of bilateral agreements can be tailored to meet the needs of both exporters and importers. The parties involved are few in number, by definition, and negotiating arrangements of mutual interest should be less complex than with multilateral approaches. We believe the type of bilateral approach outlined in this chapter has potential for improving the availability of food supplies on flexible financial terms to grain-importing developing countries.

Martin E. Abel, Susan K. Levin

5. Botswana's Food: A Case of Livestock versus Climate

Introduction

Almost all discussion on achieving food security in developing countries centers around grains. This is understandable, because most developing countries depend upon grain as their primary food source, and grains are the major food transferred among countries.

But there are many countries, or major population groups within them, that depend upon livestock as a major source of either food or income. This is especially true in certain African nations. In these lands, climate fluctuations affect livestock production as well as crop production, but the impact may be different, as are the methods available for stabilizing livestock-based food supplies.

Botswana is a case in point. It depends heavily on livestock for both food and income. It also experiences considerable climate fluctuations and instability in food supplies, and is especially subject to periodic droughts. The frequency of drought in Botswana precludes crop farming as a reliable source of food and income. With scanty and highly variable rainfall, and with drought occurring one year out of four, the majority of Botswana people are pastoralists. While livestock production is not immune from the effects of dry weather, livestock can serve as a capital resource for both food and income during drought periods, if production is accomplished under a well-managed and environmentally-sound system. This chapter, using Botswana as an example of a country dependent on livestock production, focuses on the differences involved in stabilizing and managing livestock production, compared to crop production, in situations of climatic adversity.

Sorghum and maize are the principal food crops grown in

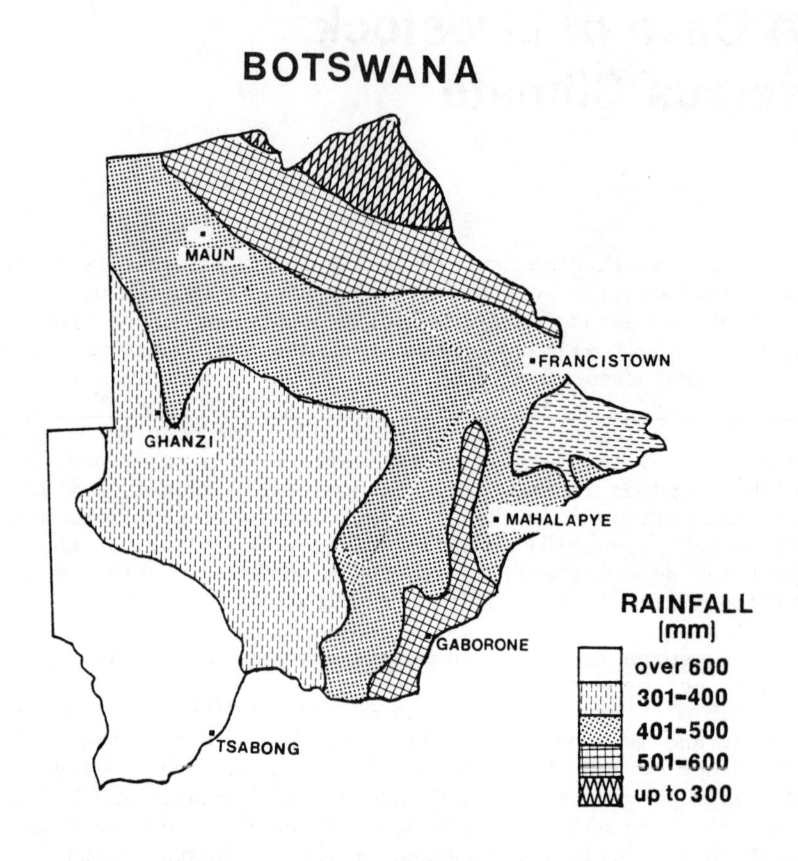

Fig. 1. Average rainfall distribution throughout the country.

Botswana. But in this subtropical south African nation, the tremendous vulnerability of crop farming compared with the greater resiliency of livestock to oft-occurring droughts, has instilled in the culture a marked preference for cattle-raising as a way of life. Even the oral literature of Botswana is filled with the apotheosis to cattle (1):

"Heavy wooden bowl of my father
When I have eaten from it, my heart is glad
For it is the bowl of my parents
Wooden bowl for sweet gravy of the cow
Lovely cattle of our home
One, alone, is sweetness
Missing one, alone, is sorrow
Dark, blue-grey cow -- one who robs of sleep
Cow with the many spots
One with the melodious tongue
Stout-one of weapons
Preparer of liquid food
God, with the moist nose."

Climate of Botswana

Botswana's subtropical climate, with temperatures ranging from minimums of 1-4.4°C from June to August to maximums of 26-32°C from October through February, is in general favorable for agricultural production. Frost may occur on 30 or so nights annually in the south, but seldom in the north. Rainfall is the limiting factor for agriculture, however. Average yearly precipitation varies between 650 mm in the extreme north to 200 mm in the extreme southwest. Ninety percent of the rain falls between November and April, usually in heavy, short showers. Except for brief periods (which are not at optimum planting times) at locations in the north and extreme southeast, potential evapotranspiration from crops exceeds available rainfall throughout the year. From one year to the next, total rainfall can vary by a factor of two. Within rainy periods, there can be gaps in precipitation as much as several weeks. At least one major gap occurs most years in every location. (Fig. 1 shows average rainfall distribution throughout the country.)

Drought (defined as a deficit between supply and requirements -- a definition that takes into account the normal usage in an area) occurs with varying frequencies, depending on location. While serious country-wide rainfall deficiencies of the kind that occurred in the mid-1960s are a rare event, coincidence of poor rainfall in two or three regions is not so rare. Weather conditions along the eastern edge of Botswana, where two-thirds of the cattle are located, coincide more than they do over the country as a whole. In the Kalahari desert, which covers a large part of the south-western portion of the country, rainfall is highly variable.

Droughts

Historically, Botswana has been subject to frequent droughts. Sandford (2) has summarized the record on drought between 1890 and 1965 based on Colonial Reports on the Bechuanaland Protectorate. This summary is presented below:

1890-'92:	A drought affected the mealie crop throughout the territory in the 1891-'92 season.
1892-'93:	Crops were considered to have failed throughout the season, due to drought and the ravages of locusts.
1894-'95:	The 1894 grain crop was a complete failure, due to a severe drought and the ravages of locusts.
1896-'97:	Rinderpest and drought followed in quick succession, and left thousands destitute. In the Southern Protectorate (now in South Africa) 95 percent of all cattle were lost. A similar situation existed in the Northern Protectorate (now Botswana).
1898-1901:	No reports.
1907-'08:	There was an almost total absence of autumn rains.
1911-'12:	The rains arrived very late and were then inadequate. Intense heat scorched the pasture land and vegetation.
1912-'13:	Severe drought conditions prevailed.
1913-'14:	The subsistence crop of maize and millet failed due to drought.
1915-'16:	The drought continued to be severe, with crops a failure in most places.
1916-'17:	Subsistence crops continued to be a failure almost everywhere because of drought.
1921-'22:	A drought during January and February 1922 meant that a shortage of grain would be experienced in the 1922-'23 season.
1922-'23:	An almost complete failure of crops sent food prices soaring.
1923-'24:	Crops, already handicapped by the unusually severe and prolonged heat of the year, and by the late and low rainfall,

were now devastated in many districts by locust invasion.

1928: Drought was prevalent in the Francistown, Tuli block, Kweneng, Ngwaketsi, Gaborone, Lobatse, Kgalagadi and Ghanzi districts.

1929: Drought conditions prevailed in many districts, especially affecting cereals.

1930: Unfavorably dry conditions in some parts affected by crops and cattle.

1931: Drought over most of the territory led to crop failures, only about one-quarter of the normal crop being reaped.

1932: Crop yields fell to well below normal because of sparse rains, and famine conditions prevailed in many parts of the Territory.

1933: The worst drought in living memory occurred which, added to a bad outbreak of foot and mouth disease, led to poverty and famine in most areas.

1935: A drought developed from the end of March which caused stock losses of up to 75 percent in some cases, and poor harvests.

1938: The area from Francistown to the south of Serowe suffered rain deficiencies which depleted harvests.

1939-'45: No reports.

1947: Drought conditions prevailed over most of the Territory. There was an almost complete crop failure and heavy mortality among livestock.

1949: Poor rains affected grazing because of the accumulation of cattle in limited grazing areas, due to the prevalence of foot and mouth disease. Crops were affected all over.

1952: Drought conditions in the south reduced crop production and led to a deterioration of grazing facilities.

1957: Drought conditions in Ghanzi and Ngamiland districts necessitated famine relief.

1959: A general lack of rain led to a poor crop season in all but the central eastern part.

1960: Severe drought conditions prevailed over much of the Territory by

	mid-winter. There was great mortality among livestock and low crop yields were experienced.
1961-'62:	A severe and prolonged drought was experienced over the Territory.
1963:	The severe drought continued.
1964:	This was the third severe drought year in succession. Farmers were unable to take advantage of the early rains because their oxen were so weak.
1965:	The drought continued.

One cannot interpret the occurrence or severity of a drought in terms of variation in rainfall alone. One also has to consider the demand for moisture, and how this changes over time. If crop or livestock production increases or moves into marginal rainfall areas, it becomes more vulnerable to shortages of moisture; thus, the incidence of drought, expressed in terms of reduced production, also increases. A recent World Meteorological Organization document (3) on drought says:

> "A most important factor in understanding drought often not included in the definitions is that it is a supply and demand phenomenon. A definition which does not include reference to water requirement or 'demand' must be regarded as inadequate. Lack of sufficient water to meet requirements is, as a definition, probably as satisfactory as any other. If the 'demand' factor is included in the definition, it follows that delineation of drought occurrence depends on the nature of the water needed. Conditions which a vegetable farmer may regard as drought, may cause a sheep farmer no concern (our emphasis). In other words, drought occurrence depends on the density and distribution of the plant, animal and human populations, their life style, and their use of the land, as much as on rainfall deficiency."

Crop Production

Grains account for about 20 percent of Botswana's marketed agricultural output and 90 percent of the Botswana diet. Sorghum is the main crop, though maize is being grown in increasing quantities. The shift to maize has become a source of concern to authorities, because it is an unreliable crop in the arid climate. A possible explanation for the movement from sorghum to maize is that sorghum requires more labor for cultivation (because of vulnerability to birds) and for milling (in order to remove the bitter seeds). Sorghum, relatively drought-resistant, is usually produced within the country, and is milled by hand in the villages (most of the maize is processed at the mill at Lobatse). Other major crops are millet, cowpeas, peanuts and tobacco. Shifting cultivation, using a plot for three or four years followed by a long fallowing period for regeneration of the soil, is the most widely practiced form of farming.

Crop production has been expanding rapidly in Botswana. Production levels, as shown in Fig. 2, are influenced greatly by variations in rainfall. Most of the increase in cereal production has come from increasing cultivated areas, while variability is due primarily to rainfall variations.

Between 1955 and 1975, Botswana's population grew from 460,000 to 692,000, or by 2.2 percent a year. Cereal production increased by five percent a year based on actual production, or by 4.7 percent based on trend production. Thus, cereal production increased more than twice as fast as population.

Livestock Production

The three principal types of livestock are cattle, sheep and goats, with cattle production being the largest. Fig. 3 shows the impact of the prolonged and severe drought of the early 1960s on livestock populations. Numbers of all types of lifestock started to decline in the late 1950s. It was not until 1966 that livestock numbers began to increase. Moreover, a strenuous effort by authorities virtually eradicated the nearly ubiquitous hoof-and-mouth disease in the mid-1960s. The country-wide effort involved policing of international borders, a rigorous cattle inspection system, and the observance of veterinary fences. An outbreak in 1978 was the first occurrence in over a decade of the scourge.

The more favorable weather of the late 1960s and early 1970s led to a substantial increase in numbers of all types of livestock. As with crops, stock production has increased at a faster rate than population. Between 1955 and 1975, the cattle population increased by 4.1 percent a year, sheep by 5.2 percent a year, and goats by 7.9 percent a year. The rates of growth were especially rapid after the major drought of the 1960s ended.

Movement of cattle to the modern, government-owned abattoir at Lobatse is facilitated mostly by railroad, though some cattle from the western regions of the country are moved on trucks. The Lobatse abattoir, the largest meat-exporting

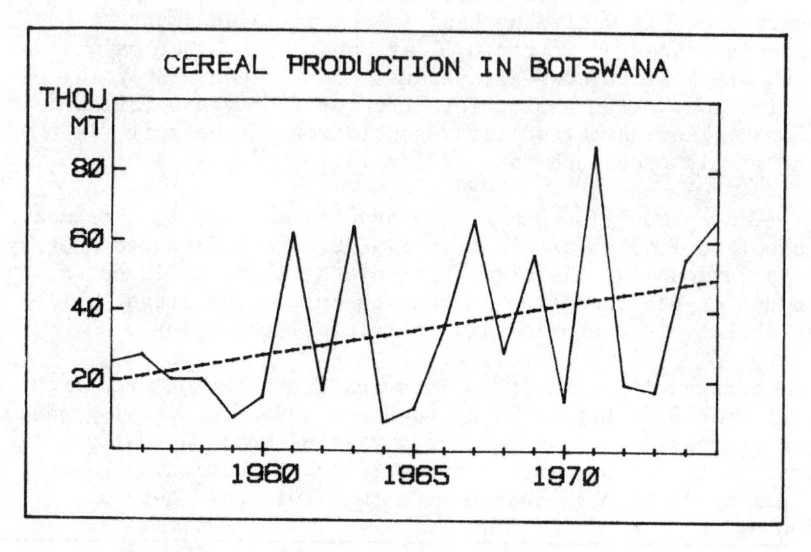

Fig. 2. Increasing cereal production in Botswana has been greatly influenced by variations in rainfall.

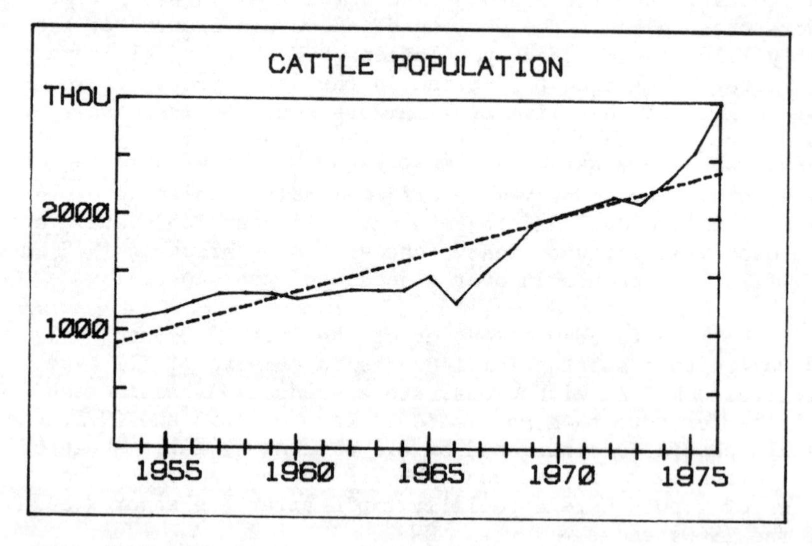

Fig. 3. Cattle population in Botswana was held down by the severe and prolonged drought of the early 1960s.

facility in Africa, has a 1,200 head/day capacity, with sales to South Africa, Zambia, Great Britain, Switzerland, Austria, and Hong Kong.

Livestock Overshadows Crops

A comparison between the livestock and crop production graphs easily reveals significantly less vulnerability of livestock to drought conditions. While both are affected by variable weather, the very nature of livestock as a capital reserve acts as a hedge against climate-related crop shortfalls.

Raising of livestock is a much more important source of income for rural people than crops. In 1974, net income from livestock was R37 million, earned by 76,500 households (80 percent of all rural households). The average annual income per household owning livestock was R481. By comparison, net income from crops was over R8 million or an average of R126 per year for households producing crops. Crops were produced by 66,000 households, or by over 70 percent of all rural households. Thus, income from livestock was nearly four times as large as that from crops in 1974.

Livestock production, the bulk earmarked for export, is more the domain of the wealthier rural people. The richest rural households derive 64 percent of their income from livestock and none from crops. For the upper-middle income group, 52 percent of their income comes from livestock and only four percent from crops. On the other hand, the poorest segment of the rural population earns five percent from livestock, but six percent from crops. However, because the poorer segments of the population work for the richer groups whose incomes are heavily dependent on livestock, the poor are more dependent on livestock for their income than is indicated by direct earnings from livestock.

Dealing with Climatic Fluctuations

Stabilizing Grain Supplies

There is little Botswana can do to stabilize grain production in the relatively low and unstable rainfall environment. Opportunities do exist for large-scale irrigation systems, although small-scale irrigation from wells is possible.

Only two strategies are available for grain stabilization. One is to store grain in years of good crops. Most rural households have traditionally done this. There may also be scope for some government storage of grain to ensure

Table 1. Cereal Supplies, Botswana, 1970-75. (In thousand metric tons; from FAO Product and Trade Yearbooks, 1975.)

Year	Production	Imports	Total Supply
1970	14	65	79
1971	87	60	147
1972	19	80	99
1973	17	85	102
1974	55	44	99
1975	66	44	110
Av. 1970-'75	43	63	106

supplies for the urban population in times of prolonged drought, but it has not been possible to determine if the government of Botswana is now doing this.

The other strategy is to rely on grain imports in years of low production, and Botswana has been doing this for some time. As a result, the total supply of cereals has been much more stable than cereal production, as can be seen from Table 1. In fact, the level of grain imports during the 1970-'75 period averaged nearly 50 percent higher than the average level of domestic production.

Livestock Stabilization

The stabilization of livestock production or supplies is a much more complex problem, mostly due to the carrying-capacity of the rangelands. The rapid growth in livestock numbers has put increasing pressure on available forage supplies. Each subsequent period of drought means that more animals, on average, are competing for a limited supply of forage. As a result, Botswana has become more "drought-prone," using the World Meteorological Organization's definition of drought based on a supply-demand concept. Also, as grazing pressure intensifies, the amount of "permanent" damage from overgrazing increases with drought. In order to survive, animals increasingly overgraze during drought periods, and may cause permanent damage to the natural vegetation. When the rains return, the natural vegetation recovers more slowly, and may not reach its previous level of productivity, thereby making the livestock population even more vulnerable to future droughts.

Sandford has computed a range condition index for the 1925-'75 period, shown in Table 2. The drought of the early 1960s was especially severe and prolonged in three areas -- Francistown, Mahalapye, and Gabarone -- while it was more moderate in the other three areas. It is unusual for a severe drought to occur simultaneously in so many regions. During this period livestock numbers were reduced. The range index also shows that conditions have been unusually favorable in the 1970s, which helps to explain the rapid growth of livestock to record numbers in this period. It will be more difficult, assuredly, to sustain these large numbers of animals during the next drought than it was during previous ones.

While it is not possible in a country like Botswana to completely stabilize livestock supplies in the face of climate fluctuations, it is possible to evolve strategies that will contain variations in livestock numbers. The first

Table 2. Botswana Range Condition Index: Percentage Deviation from Median Rainfall. (From Sandford, S., Dealing with Drought and Livestock in Botswana. A Report to the Government of Botswana, May, 1977.)

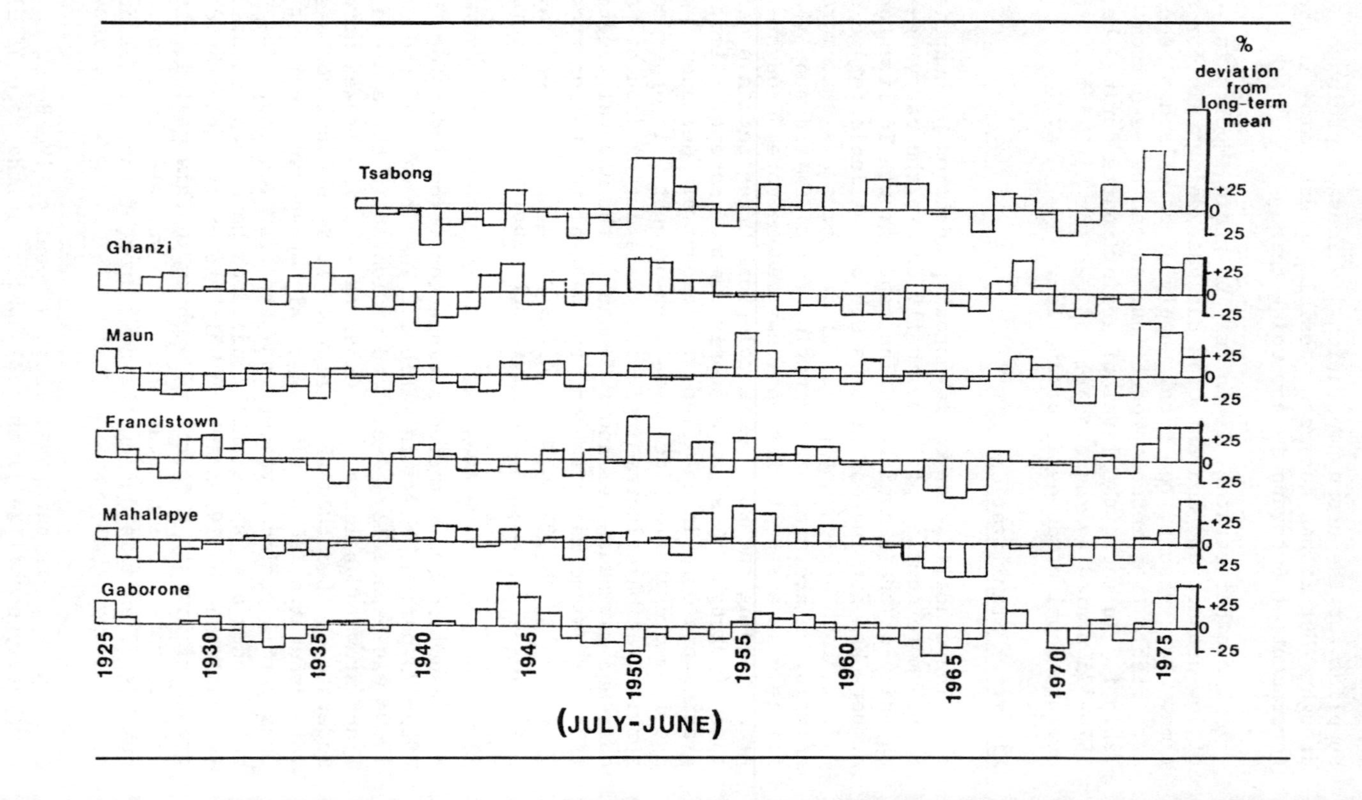

element of a drought strategy is to prevent livestock numbers from increasing rapidly in relation to forage and water supplies. Unless this is done, the loss of livestock numbers and the incidence of permanent damage to the forage supply will become increasingly severe with subsequent droughts.

Sandford recommends encouraging a relatively stable livestock population over time, through the timely slaughter of animals when drought occurs. This approach requires adequate slaughter capacity and transportation, as well as economic policies that favor herd liquidation during drought. Since Botswana is a meat exporter, increased meat supplies during periods of accelerated slaughter would have a ready market.

Sandford lists the following elements of a strategy to contain livestock numbers (4):

1) Legal or taxation measures to force livestock owners to destock;
2) The provision of special prices or other attractions to provide a good incentive for farmers to sell early during drought, at a time when marketing or processing facilities are under-used.
3) Guarantees to farmers that they will be able to purchase breeding stock at reasonable prices at the end of the drought.
4) The provision of feed lots at the slaughterhouse whereby to put animals who are in very poor condition into better condition for slaughter.
5) The provision of adequate capacity in the marketing and processing system to cope with peak flows in time of drought. This capacity may include:
    a) Provision of extra finance to existing buyers, and additional buying posts in cases where existing buyers cannot cope.
    b) Adequate capacity for veterinary inspection and quarantine.
    c) Sensible allocation of quotas to maximize offtake from most seriously affected areas.
    d) Adequate transport facilities, including rail, trucks and lorries, adequate water and grazing on trek routes, plus additional feed where adequate grazing cannot be got; adequate holding grounds at railhead and abattoir in order to facilitate a smooth flow of cattle into slaughter.

e) Adequate processing and storage facilities including: adequate capacity at Lobatse or other major abattoir; salvage facilities in producing areas to process animals which will not survive as far as major abattoirs, so as to salvage their hides and by-products.

f) Adequate external markets to take an increased flow in time of drought.

A drought information system is also suggested to provide timely information on forage and animal conditions in each region. Such an information system would help plan for liquidation of animals, for movement of animals from drought-affected areas to those with more adequate forage supplies, and for the provision of supplemental feed.

While the ability to provide supplemental feed is limited, it can nonetheless be important. Hay-making is possible in parts of Botswana, and this hay can be a forage reserve in times of drought. It would not be sufficient to take care of all animals, but it would provide a source of feed for work stock and breeding animals. However, haying is relatively expensive, and only the large (rich) farmers could afford it. To ease the burden of drought on the poor, for which preservation of work stock is vital, the government might maintain a hay reserve.

Another measure could be the planting of spineless cactus, which grows well in Botswana. This is used as an animal feed and source of water in times of drought in many other countries. It represents a "natural" feed reserve.

Feed concentrates could also be provided in times of drought. But these are expensive, since they are based on grain or oilseed products, or consist of high-quality forage such as alfalfa cubes. In addition, feed concentrates would have to be imported. But such concentrates could be used to ensure survival of work animals and breeding stock.

Water wells could also be provided for use only during drought conditions. One reason for the rapid growth in livestock numbers and the higher grazing intensity in Botswana has been the expansion of wells. The water supply has increased faster than the supply of forage. This practice, common in many other African countries, should not be encouraged. But it may make sense to have wells available to use during droughts, because they may be necessary to ensure the survival of the desired number of animals. They could also be an important adjunct to drought-induced migration to areas with more abundant forage supplies.

Finally, there are areas in Botswana that are infested with tsetse flies and not usable for livestock production on a regular basis. These areas are normally swampy wildlife preserves. As an emergency measure during drought, Sandford recommends considering the temporary eradication of tsetse flies through spraying, and the movement of livestock into these areas. While it is not economical to maintain a tsetse eradication program on an ongoing basis, it may be justified in times of drought emergency.

Conclusion

As has been indicated, extreme variability in rainfall and chronic drought that make crop production highly unstable may not affect livestock production to the same degree. Hence, in a country like Botswana, livestock production is a major source of food and income for most of the population, and may provide a hedge against food insecurity in time of crop shortfalls.

The primary problems in dealing with this dependence on livestock production are to manage livestock numbers, both during droughts and otherwise, so as to avoid permanent damage to forage supplies when the droughts do occur, and to provide supplemental feed during periods of severe moisture shortages. Without constant adequate forage, livestock production could not exist, and Botswana would suffer periodic serious deficiencies in food supplies and income, and would have to rely increasingly on food imports.

The strategies for managing livestock numbers that were described above are not the only ones, but they do indicate the nature of possible solutions to the problem of countries dependent on livestock production in the face of climatic instability.

References

1. Hartland-Thunberg, Penelope. Botswana: An African Growth Economy. Westview Special Studies on Africa, Boulder, 1978.
2. Sandford, Stephen. Dealing with Drought and Livestock in Botswana. A report prepared for the Government of Botswana, May 1977.
3. World Meteorological Organization. Drought. Special Environmental Report No. 5 (WMO No. 403), Geneva, 1975.
4. Sandford, op. cit.

David P. Harmon, Jr.

6. Case Studies of the Multinational Corporation as a Food/Climate Buffer

Introduction

Undesirable climate and runs of bad weather, since they cannot be directly controlled nor confidently forecast within a narrow range, are usually the most critical influences on agricultural success in developing countries. It thus behooves farmers and governments to insert stabilizing factors or "buffers" in the food/climate system to protect against its inherent year-to-year and within-year variability. Multinational corporations (MNCs), by creating the buffering action of food processing and marketing resources, play an important role in reducing the vulnerability of a developing country to climate impact. This cushioning effect is also important because it can influence the more rapid adoption of promising new technologies. In an area of high weather/climate risk, a farmer will be very cautious in adopting new technologies because of the perceived risk of failure and difficulties of assessing benefits and costs. Hence the MNC, if accepted and used advantageously by the developing country, can not only be an agent of growth and provider of technology, but also a useful buffer in the food/climate system. It follows that the policies that are adopted by developing countries to provide an economic and political atmosphere conducive to multinational participation in agricultural and economic development, will be a major issue of the 1980s. Two case studies of MNC activity in Chile and Algeria are presented to explore this issue and interaction.

This paper has been adapted from a chapter, "The Multinational Corporation: A Buffer in the Food-Climate System," in Critical Food Issues of the Eighties, Marilyn Chou and David P. Harmon, eds., Pergamon Press, 1979.

Stages in Agricultural Development

The solutions to tomorrow's "food problem" must come in large part from the problem countries themselves. While the availability of resources and technology over the long term leaves little doubt as to the potential of both conventional and unconventional agriculture, achieving this potential is far from inevitable. Necessary for the task of feeding a growing population is the ability of individual governments to marshall their resources, and to design and implement the appropriate policies to capitalize on them.

As developing countries try to increase their agricultural productivity today, they benefit from existing technology. What has taken years of research and development, and sizeable investment by one nation or research center, is available for their use and adaptation. Today we tend to transfer with only moderate success the skills and techniques that have been developed in the expansion of temperate-country agriculture. In the future, with adequate infrastructure and institutions in tropical climates, and improved ability to adapt agricultural technology to specific area conditions, there should be vastly accelerated transfer of temperate-country technology to the developing world. Once institutional mechanisms are established to facilitate this transfer, developing-country farmers will be able to increase their productivity at lower costs, and at much faster rates than the farmers in countries who had to cope with the problems of the new technologies and refine and adapt them to their needs.

In the early stages of agricultural development, a country is especially vulnerable to climatic fluctuations. It lacks the productive capacity, grain reserves, complex infrastructure and other buffers to tide it over periods of shortage. At this point, backstopping in the form of food aid is critical for that country. Also, assistance of the type that will help the private sector grow is fundamental to development. A fair degree of political and economic stability is necessary for MNC participation, including the transfer of skills and technologies.

Sidestepping with Technology

Insurmountable institutional obstacles may prevent a country from developing the necessary infrastructure for agricultural development. They may include not only physical and financial deterrants, but also resistance to change from those who stand to lose from it. Climatic obstacles may also exist, such as the failure of the monsoon two years out of five in India.

When a nation faces serious institutional and/or climatic obstacles hindering agricultural development, what are the alternatives? Technology offers peoples and governments the means to achieve at least some of the pressing objectives by sidestepping, rather than meeting head-on, some of the more serious institutional obstacles.

Technology offers alternatives at three important points in the field-to-table food chain. The first point is on the farm, and ranges from on-the-shelf technologies and agricultural sub-systems (water and land management, mechanization, cultural practices, improved livestock productivity, etc.) to mission-oriented basic research. MNC involvement in the former is more likely, while government, foundation and quasi-public organizations' support is vital for the latter. In the long term, mission-oriented basic research holds great promise in three areas: photosynthetic efficiency, biological nitrogen fixation and genetic improvement. Better "harvesting of the sun" and subsequent partitioning into the plant's harvested parts, enhancement of biological nitrogen fixation by legumes, and the genetic improvement of plants are of great interest because they represent ways to greatly increase agricultural production, yet at the same time are apolitical, scale neutral, non-polluting and easily transferable.

The second point occurs after food is harvested, and entails the technologies of loss prevention (of both quantity and quality) at numerous locations from field to table; viz, storage, in-transit, processing, packaging and distribution. Post-harvest losses worldwide are estimated at 10-20 percent of the harvested crops, and it is reasonable to assume that such losses are higher in the developing world. Loss prevention technologies are available and easily adaptable, especially as a country's industrial sector modernizes. They are apolitical, but entail capital, management expertise, skilled personnel and training -- areas in which MNCs excel. Better storage of locally-produced foods increases food security, gives greater incentive to produce, and draws more rural people into the market system. A three or four day extension in the shelf life of perishables without refrigeration increases the availability of food.

The third point occurs in the processing stage, where food can be fortified with nutrients that may be lacking. The fortification of common foods such as bread, fats, salt, tea and cereals, offers an inexpensive and rapid way to reach large numbers of people without the massive socioeconomic changes that happen when eating habits change. It also offsets the nutritional loss that tends to occur in the early stages of economic development as people switch to aesthetically more pleasing, but often less nutritious foods. Nutrient fortification involves only modest changes in food processing, no alteration of the organoleptic qualities of foods, nor or sociocultural habit -- eating, cooking or buying -- and is especially suitable for large concentrations of people with nutritional deficiencies (1). It is attractive because of its relative simplicity and low cost, and is perhaps the most important nutritional advance that can be given to the developing world. Fortification technology, an important area for MNC activity, has been largely responsible for the disappearance of pellagra, rickets, ariboflavinosis, and beri-beri in the United States.

Of the principal factors of production, technology is special in that it not only has a heavy short-term impact on productivity, but also has the attribute of rapid mobility. To be applied directly, however, it must be commercialized or adapted into a useful product or productive capability.

The MNC's Technology Mission

A mission of the MNC is to transfer and commercialize the technologies it creates. It is the most important institution with a global capacity to apply science and technology to both production and marketing, and to transfer technology across national borders (2). Technology is transferred by the MNC in simple to rather sophisticated forms: 1) Products; 2) Technical assistance to customers, users and local suppliers of materials and components; 3) Local service facilities; 4) Local production facilities and associated training of personnel, and 5) Long-term arrangements ("product-in-hand" type contracts) including (4) plus organizational structure, management expertise, internal accounting systems and access to new technological developments over a specified period of time.

The MNC is indispensable for the agricultural and industrial development of later-comer nations. There is such a rich flow of technology across borders that the "wise" developing country should design its trade and investment policies so as to exploit this flow, rather than keep out foreign technology or duplicate what has been or is being done in the developed world (3). Developing countries are simply unable to generate enough scientists of sufficiently high quality themselves. Moreover, rapid adaptation to change is something that business does well. In a competitive market environment the reward/penalty "system" ensures efforts to exploit technologies. The need for the MNC as the agent of technology transfer to the developing world is perhaps best expressed by the words of Fernando Monckeberg, Chile's formost nutritionist:

> "When the scientific and technological level is low, so is the educational level, the organization of the community is inefficient and productivity is low, all of which leads to stagnation and poverty." (4)

Following are two case studies, the first of which presents a novel, effective means of MNC-developing country cooperation. The second portrays a country with many of the problems and constraints the MNC faces in the developing world.

Case Study 1: Fundacion Chile

Too often, governments discourage agricultural development by establishing ill-conceived policies. Many countries prematurely emphasize heavy industry, thus causing a misallocation of resources in the economy, a patchwork of policies for the agricultural sector, and improper integration of the agricultural sector into the overall development framework. Moreover, there is the lack of an adequate theory of the relationship of institutions to technological change. Public institutions are thus largely unable to respond to the substantial social benefits inherent in new technologies (5). This inability to respond arises in large part from:

- The lack of an effective organization to produce a flow of highly profitable innovations suited to local conditions;
- The failure to understand that agricultural development is an iterative process in which the government planner must constantly look for new limiting factors;
- An excessive emphasis on the public sector's fulfillment of development functions, which stifles the private sector;
- An excessive centralization in decision-making and a shortage of qualified personnel to handle highly pyramided operations (6).

Often the private sector is not permitted to operate in those areas where it is inherently more efficient than the public sector. Further, the proper role of the public sector vis-a-vis the private sector is not recognized. The public sector should provide those inputs the private sector cannot -- because not enough of the benefits can be captured by the private sector to justify investment. Mission-oriented basic research is a good example of the kind of input the public sector should supply.

In this case study we see how Chile and International Telephone and Telegraph (ITT) collaborated to establish an "institutional mechanism" that permits agricultural, nutrition and telecommunication technologies to be taken up and adapted to Chilean conditions. ITT and the Chilean government in 1974 jointly established and funded Fundacion Chile, a "laboratory" designed to facilitate the transfer of technologies to Chile. It is not, however, a research center in the usual sense of the word. Most developing countries have research centers, but they lack the ability to transfer knowledge generated at the indigenous research center to the factory and to the marketplace. Fundacion Chile's objective is to meet and overcome this lack. Moreover, its technological assistance is directed at Chile's locally-owned and run businessess -- to increase productive employment and to develop local and export markets.

Beyond the Market Factor

Under normal circumstances the multinational firm identifies and selects favorable marketing opportunities. Engineering, production and marketing capabilities are built up in the chosen markets. Fundacion Chile goes three steps further, however -- and these are the key to helping locally owned businesses assimilate technologies and prosper. The first step is improving industry-government communications. Poor industry-government communications are often the rule in developing countries, because of the preponderance of military governments. Differences in background, experience and interest give rise to uncertainty surrounding government intentions and policies regarding business. The result is a conservative risk-avoiding approach on the part of businessmen. Fundacion Chile formed business advisory boards on which local agro-businessmen, customers and government staff participated in advising the foundation and in listening to one another's problems. Workshops on major (and general) concerns, such as production, quality control, packaging, cold storage and export market planning followed. Out of the workshops came follow-up plant assistance programs.

The second step involved overcoming the basic conservatism of the local investor and the local manager, and the difficulty of raising capital in the face of high rates of inflation and typically high debt-equity ratios. Fundacion Chile searches out programs that minimize new plant and

and capital requirements, that utilize idle plant capacity, and that improve the efficiency of local resources and people utilization. Investment analysis is carried out to establish credibility with local sources of funding, to enhance profit margins, and to permit self-financing for modernization and expansion.

The third, and most critical step, is the development of market intelligence. The lack of hard marketing data in Chile is overcome by teams of Chilean economists and multinational marketers who define markets, competition, distribution channels, and needed plant, equipment and material resources.

A Technology Assistance Center, divided into two sections, was formed. The first section deals with market and business development, prepares investment analyses of proposed projects, and incorporates an early warning system which alerts the participants to project problems through customer contacts, plant visits and market surveys. The second section is divided into technical departments along the lines of Chilean agro and marine business, i.e. food and feed grains, fruits and vegetables, dairy products and marine foods. A sampling of projects undertaken follows:

1. Development of high nutrition baked goods using domestic grains;
2. Cold storage for vegetables;
3. Fresh vegetable and fruit product development for export markets;
4. Sun-drying of apricots, raisins and other fruits;
5. Strawberry production;
6. Dehydration technology for apples, green peppers and garlic;
7. Development of shell fish seedling for local maricultures;
8. Clam harvesting;
9. Training courses in food-processing quality control and plant sanitation to overcome food losses (7).

Furthermore, the center operates these projects in a "team" fashion with various Chilean institutions (universities, businesses, cooperatives, government laboratories, etc.) so that management skills and technology pass to Chileans.

Self-Sufficiency: The Goal

The underlying objective of the foundation is to improve

Chile's ability to feed itself and to provide export food items with a high proportion of local contributed value. By working with the Chilean federal nutrition organization, CONPAN, the foundation is keyed into those population segments where food and nutritional needs are most pressing. Two projects recently carried out must be noted because they are indicative of the ability of technology to blunt climate-imposed variations in food supply and side-step institutional obstacles to increased food supply. For many years, Chile has provided centrally-purchased, but locally-prepared food to poor school children. The system, however, was crude, unsanitary, and wasteful. The foundation, in collaboration with CONPAN, developed an inexpensive, concentrated processed food biscuit which is manufactured by a local baker. The biscuit is high in protein, is fortified with vitamins and minerals, is sanitary, has a long shelf life, and most important of all, is liked by the school children. Chilean ingredients, Chilean technologists, a Chilean biscuit manufacturer and American technology (ITT/Continental Baking) adapted to Chilean requirements met an ultimate need, sidestepped "institutional" obstacles and put a "buffer" in the food-climate system. The second project reduced post-harvest food losses for various fruits and vegetables in the distribution chain -- by concentrating on post-harvest physiology, food-processing plant quality control, sanitation, new product development and dehydration techniques.

Fundacion Chile is an experiment in providing the "mechanism" that will allow both the public and private sectors to respond to the benefits of technology. It is equally funded by ITT and the Chilean government, which insures the continued interest of the latter. Such an organization, by drawing on the many talents of one of the largest multinationals, can be the catalyst for increased food production, better nutrition, and less susceptibility to climate-induced food production variations. All of Latin America is watching Fundacion Chile with interest.

Case Study 2: The Algerian MNC Crucible

Algeria is a developing country whose agricultural production is constrained by climate, and whose agricultural and industrial sector development has been limited by both political philosophy and institutional obstacles. For purposes of shaping a strategy to stimulate MNC participation in developing countries' food and agriculture sectors, and to transfer technology, Algeria is an ideal country to consider because it is:

1. Young, outwardly socialistic, but basically

nationalistic, with a heavy residue of "free enterprise spirit" among the people;

2. Politically left on the world scene; fairly radical in the Arab world; trying to take on the leadership of the Third World;

3. A nation with real potential for development:
   a. large national gas reserves; significant oil reserves
   b. eight-nine month growing season
   c. hardworking, relatively capable labor force
   d. potentially large supplies of non-fuel raw materials
   e. a growing economic infrastructure

4. An extremely difficult place in which to operate -- even as an Algerian organization:
   a. differences in business practices, "mores," and ethics
   b. deep suspicion of outsiders
   c. shifting requirements for foreign companies planning to, or doing business in Algeria
   d. joint ventures possible (51 percent or more Algerian ownership mandatory) but with constant threat of nationalization.

Algeria gained its independence from France in 1962 after an eight-year-long struggle, in which one million people were killed. As one might imagine, it was flat on its back economically. Ninety percent of its fifteen million people live along the littoral, which supports the bulk of Algeria's agriculture, with the exception of wheat and sugar beets. Its principal food grain crop, wheat, is susceptible to drought, desertification, and to the sirocco -- the hot, dry wind that blows north off the Sahara just before harvest time.

It is a country of contradictions: suspicious of outsiders (sometimes to the point of paranoia) and politically hostile to the United States, but aware of the benefits MNCs can bring, and relatively eager to do business with U.S. firms; nationalistic, but filled with Arab-Berber factions and jealousies; and socialistic in government, but with a population that is basically capitalistic. It is one of the most difficult countries in which an MNC might do business, not the least reason of which is the European/American difficulty in understanding the Arab mind, and vice-versa. Because of these contradictions and the difficulties they imply, it is a crucible for those MNCs which operate there, and good lessons can be drawn from the "Algerian experience." Most MNC dealings are either with Algeria's national companies or with various ministries/agencies of the government.

Productive Goals Plus Bottlenecks

Ten years ago, Algeria embarked on a real economic gamble -- that of rapidly starting many technologically complex and capital-intensive industries, as well as a viable economic infrastructure. In the founding period of the national companies, production was stressed to the exclusion of all else. Today, other functions (planning, marketing, distribution, management, personnel) are being recognized, but serious bottlenecks exist in doing business.

The public industrial sector suffers from serious supply and distribution bottlenecks. These are largely due to the administrative separation by the government of business functions which properly belong to the national companies. For example, one ministry is responsible for the supply of raw materials, another for transportation, and a third for marketing final products.

Other bottlenecks arise through lack of coordination between agricultural agencies and particular national companies; in some cases, other ministries' agencies and national companies do not even communicate with one another, when in fact, the closest of cooperation is vital. The necessity for cooperation is evident, when one understands that Algeria often has as little as ten days of wheat reserves in storage and suffers from extreme port congestion. Because of climatic and political conditions, her wheat crop does not meet its potential one year out of two. The power accorded various ministries and agencies to determine prices, terms of payment, quantities and qualities of goods which they will purchase from some of the national companies has reduced particular companies to the role of production entities, rather than companies responsible for the entire spectrum of normal business activities. This emasculation means that these companies cannot function efficiently, that discord and friction arise between the company and agency/ministry, and that the economy suffers.

Long-Range Planning and Investment

In many national companies, as well as in many ministries, there is a severe lack of long-range planning. Lack of planning, failure to consider alternatives, and a lack of broad view, cause misplacement of investment funds through faulty investment decisions. For example, factories are often built to replace old plants, when in fact the old plants should merely be shut down.

In the past ten years the economy has grown in size and complexity to the point where administrative and structural bottlenecks can strangle, and investment mistakes can have very serious repercussions. There is a critical need not only for "strategic" planning but also for "tactical" planning, and redefinition of areas of responsibilities.

This then, is the business climate in which an MNC must operate, whether it is only a supplier of goods or a 49 percent partner in a joint venture with a national company.

The foregoing background comments on Algeria's economy are essential to understanding the need for technology appropriate to a country's (and its various sectors) stages of development, both present and future. In the 11 years between independence and 1973, Algeria rose from a devastated country to one having to come to grips with some very sophisticated questions; viz, a development policy of industrialization or of agricultural sector-led growth; capital rationing; the acquisition of technology; meeting the nation's current and future food and nutritional requirements in the face of severe climatic difficulties. Examples of investment failure abound.

The Superamine Story

A poignant example of failure was the constuction and operation of a plant to turn out a high vegetable protein food, Superamine, comprised of chick peas, lentils, soya and wheat. While the idea of increasing the availability of proteins via locally-grown products, with foreign technology, was certainly laudable, poor planning, financial constraints and severe contamination problems led to the project's demise. On the other hand, three industrial flour mills (to be located in the East, Center and West of Algeria's littoral -- covering 90 percent of her population) have been in the planning stage for the past ten years! Algeria is a nation of bread and couscous eaters -- both wheat products -- most of which she grows herself. In this case, however, technology offers a low-cost effective solution to much of Algeria's nutritional deficiencies, and secondarily places a needed buffer in the climate-food system.

Since nutrition, like agriculture, is "location-specific," fortification of wheat flour, a food eaten regularly and in consistent quantities throughout the year by all age and economic groups, could help achieve Algeria's nutritional objectives.*

The Sonacome Success

Technology transfer entails more than just the transplanting of equipment, a plant or a technology itself. It involves planning, managerial expertise, cost accounting and measures of performance, education, and understanding that development is both a dynamic and iterative process.

The success side is perhaps best typified by Algeria's automotive industry. Sonacome, the national automotive company, identified important interrelationships among various national needs, as well as among company needs. Sonacome's tractor-producing facility at Constantine not only satisfies Algeria's need for tractors, but has also given Algeria expertise in manufacturing diesel engines, which she will produce in greater quantities for the buses, trucks, vans and industrial vehicles to be built at other Sonacome facilities.

Having run its first fully-integrated operation, Sonacome can transfer this expertise to its planned automobile assembly facility. The Constantine tractor plant is a self-sufficient source of supply for all diesel motors in Algeria, and coincidentally, has met Algeria's need for tractors. Sonacome employed the "produit-en-main" (literally translated, product in hand) concept of acquiring foreign technology. Sonacome displayed forward thinking, because not only did this strategy entail the construction of the tractor plant and associated training of technical and work staffs, but it also involved the acquisition of a management organization, internal accounting and performance measures, a guarantee of output under full operating conditions and a ten-year access to technological innovations in all aspects of tractor design and construction from the constructor firm, DIAG (West Germany). Sonacome demonstrated that it understood how to respond to, and take advantage of, the benefits of technology. The tractors produced are simple, unadorned, durable and appropriate to Algeria's stage of agricultural development. They are also suitable (and exportable) for other countries of the Maghreb and the Middle East.

In general, there is a high degree of political and eco-

---

*Industrial countries take fortification of foods for granted. Their citizens assume that salt is iodized and bread and milk fortified. This is a case of an on-the-shelf technology that is directly transferable from developed to developing countries.

nomic risk in Algeria for the MNC. At best, most firms will go no further than turn-key projects, and much prefer straight sales of equipment and production lines for hard currency payable elsewhere than in Algeria. Thus, it is not surprising that there are very few joint ventures operative in Algeria, and little MNC activity in the agricultural sector, where long-term commitment is necessary. The country just poses too many obstacles and risks. Its political and economic environment for MNCs, however, is similar to that of other developing countries, and is therefore a good springboard into the development of appropriate MNC developing-country strategies to improve agricultural and food sector development and to speed the transfer of technology, thereby reducing climate-related vulnerabilities.

The Problems of MNC Participation

With the foregoing case studies as references, let us now consider some of the problems which influence the success of MNC participation in agriculture and food-sector development.

First, there is the problem of lack of incentives and LDC fears. In the August 1977 issue of Fortune magazine, Sanford Rose, author of the article "Why the Multinational Tide is Ebbing," gave two views of the MNC -- "new Bully on the World Block," and "one of the greatest forces for progress yet devised by man." Whichever view is ascribed to, it is clear that there is increasing MNC reluctance to move into any but the most stable of developing-country markets. While MNCs have little security of tenure in the developing world, the U.S. Congress has whittled away at their tax status. Between 1971 and 1975, in low technology and in highly competitive areas, approximately 10 percent of all U.S. foreign subsidiaries were sold off (8). There is little commercial bank funding of projects to increase agricultural production in developing countries, because earlier loans for similar projects have been found to be shaky, and the projects lacking in both management and technical feasibility (9). Multinational firms, too, are often loath to get involved in projects to increase food because of past bad experiences -- many with formulated protein-rich foods. They were simply too costly for the market, and in many cases the companies involved did not recoup their investments, much less make a profit.

MNCs face many restraints in developing countries, although since the oil crisis some capital-poor LDCs have softened their opposition. The restraints range from control over allowable license fees, royalties, dividends and branch earnings a subsidiary may remit to its parent, to outright expropriation. One insidious form of restraint is the process of "unbundling," whereby the developing country seeks technologies through licensing, financing through international banking channels or domestic savings, and management expertise locally or from abroad. Thus, the developing country attempts to get the benefits of the MNCs without actually admitting them to the country. Another form of restraint is that of creeping expropriation, or the "obsolescing bargain," where the MNC is forced to give up increasing amounts of control once the investment has been made (10).

On the other hand, how are MNCs viewed by developing countries? Many governments are uneasy about MNCs, because they are unable to predict (and control) their behavior -- even if the subsidiary is completely willing to adhere to the laws of the particular country. Their unease also stems from the following:

1. Some MNCs have more economic power than the developing country itself.

2. Short-term capital movements by MNCs can create balance of payments problems for a country. For example, the timing of payment of intercompany accounts can be adjusted to take advantage of potential currency devaluation.

3. Political pressures, fears of neocolonialism, North-South frictions, developed country trade restrictions on developing-country manufactured goods, add to the uneasiness. (11).

Steps Towards Synergism

Mutual understanding of each other's problems and objectives will lead to much more productive interaction between the MNC and a host country. A first step is in incentives for the MNC and a spirit of "pro bono publico." The developing country must understand that MNCs simply will not operate in situations where risks and restrictions, both commercial and political, are too high. Second, MNCs can only make a difference in food supply in money economies. Third, developing countries must recognize that private enterprise is and will be the predominant creator and transferrer of industrial technology. Developing-country public institutions are only capable of modest contributions. The specific measures that the developing country can take to create an environment conducive to MNC participation start with the developing country's knowing its needs and making them known to potential foreign investors through planning and enabling legislation. Other measures are the classic ones of investment guarantees, tax concessions, reasonable currency repatriation allowances, a viable legal system -- in short, fundamental and permanent assurances of a reasonable business climate in order to obtain foreign capital and technology (12).

The MNC, for its part, could enhance its chances of long run success by adopting a spirit of "pro bono publico" in its initial endeavors in developing countries. The MNC and the host government must work in complementary fashion, with specific agribusiness and food industry projects fitting into overall government programs. The MNC can take a portion of the "backstopping role" that it normally ascribed to foreign grain-exporting countries -- usually the United States. It can do this by using its "comparative advantage" in transferring resources (capital, management expertise, and technology) and in commercializing technology.

A second step is to foster a positive developing country response to technology -- with the MNC as the mechanism. While there are several organizations, some private and some quasi-private, which act as "synchronizers" of agribusiness and governments in the developmental process, little attention has been given to the possibility of the MNC as the "mechanism" which could enhance public sector response to technology. As indicated earlier, the MNC as an agent of change is eminently qualified to react rapidly to change and is highly capable of transferring and communicating technology and expertise. What are the general requirements?

1. Both the MNC and the government should share equally in the operation of <u>and</u> the funding of the "institutional mechanism." Funding by the government is critical in order that the government have a stake (which it could lose) in the institution. That is to say that both parties share the risk and rewards equally.

2. Recognition that most of those people suffering from malnutrition will only benefit from expanded food production if they are brought into the money economy.

3. To as great a degree as possible, local resources should be used to fulfill local needs. This requirement is consonant with developing country desires to keep as much as possible of the "value-stream" of a product in the country.

4. Technologies appropriate to the country's/sector's stage of development should be used. Too often developing country planners want to "catch-up" immediately and are mesmerized by economies of scale.

5. Recognition that food, from production to diet, is a dynamic area, and therefore requires flexibility in policy and project planning and implementation.

6. Realistic evaluation of a country's current and likely future agricultural production system cannot be bought, but rather it must be developed -- as part of a country's overall economic development.

7. Recognition of the developing country's need to develop its own technology if it is to be competitive in domestic as well as in international markets.

8. That government-multinational collaboration enable a domestic scientific/technical infrastructure capable of generating knowledge.

What are the Specific Requirements?

To meet the objective of blunting climate-induced variability in food supplies, developing country-MNC collaboration should fulfill certain requirements. These are:

1. Projects to increase food production and food processing must produce for the marketplace -- which means price incentives must be present. Farmers and MNCs alike need economic incentives to increase production. Moreover, such projects translate into employment and value added.

2. Maximum advantage of science and technology must be taken in the design of projects -- especially where the country is poor in natural resources.

3. More than one crop/one enterprise must be emphasized, in order to cushion against adverse crop developments.

4. Existing infrastructure, especially water supplies, communications, and transportation/distribution networks must be improved.

5. Market opportunities, both domestic and foreign, must be clearly defined.

6. Existing technologies should be tailored to the country's specific requirements and its comparative advantages -- raw materials, labor skills, natural endowment, etc. This must be done with care so that the MNC is not open to the charge of having provided second-rate technologies (13).

The area in which the MNC can have the greatest impact is in the provision of a reliable, waste-minimizing distribution system from the farmer to the consumer. Both in post-harvest food conservation and in human nutrition, the MNC can truly be a "buffer" to climate-caused variations in food supply.

Effective central storage of grains is one of the most important ways a country can protect itself against fluctuations in agricultural production. Even though the required capital investment is high, the social returns are high and the facilities are long-lived. Moreover, the need for improved handling and distribution techniques runs all along the farmer-to-consumer food chain. No organization is better versed in training and application of safety, food purity, and sanitation standards in food processing than a multinational food corporation. Another area that is a "natural" for the American MNC is food research, since much of this work is done in the private sector in the United States. Further along the food chain, a particularly effective area in which effort should be placed is in food service for the developing country's institutional markets. By concentrating on providing a balanced diet from local foods, and on decreasing food waste, pressure can be placed on the food chain from farmer to processor to supply foods of requisite quality in required quantities.

Conclusion

The MNC is one of the most powerful agents of change, transferrers of technology and potential creators of income. All three facets are important points of intersection of the developing country and the MNC. If a foreign subsidiary is successfully established, it becomes an appropriate conduit for the transfer of innovations. Moreover, it is part of the international marketing network of the MNC, a network which is extremely difficult to duplicate (14).

For a multinational to enter a developing country and develop a viable business, the climate -- economic, social political -- must be conducive to meeting the objectives of both the country and the multinational. Laws and regulations affecting foreign investment must be reasonable and stable, and there must be a market for the MNC to serve. Moreover, the developing country must be committed to the success of projects entailing MNC participation.

For its part, the MNC must be as apolitical as possible (15), be willing to accept a minority position in created subsidiaries, provided the returns are adequate to justify investment, be prepared to reinvest earnings locally, and even be willing to shift part or all of its equity to local ownership, as the business and local capital markets evolve (16).

Willingness to accept a minority equity position or even non-equity arrangements such as management contracts, production sharing contracts, and technical assistance arrangements is necessary because relative bargaining strengths are shifting dramatically from MNCs to the developing countries. In short, MNCs must maintain their "legitimacy" in host countries. Legitimacy means contributing to host country values and objectives. This is especially true in developing countries that are in the intermediate stage of development -- are developing their own capabilities and thus "value" the MNC less.

In its normal areas of activity, production, marketing and transferring technology and expertise, the MNC can bring significant opportunities to the agricultural and food sectors of developing countries. These opportunities can help protect the developing country against the vicissitudes of climate. Enlightened developing country leadership will take the words of Fernando Monckeberg seriously:

> "Changes are happening too fast -- to ignore them, to try to change their nature or to minimize them means missing opportunities that may not come again." (17)

References

1. Berg, Alan. The Nutrition Factor -- Its Role in National Development. The Brookings Institution, Washington, D.C., 1973.
2. Poats, Rutherford M. Technology for Developing Nations. The Brookings Institution, Washington, D.C., 1972.
3. Ibid.
4. Monckeberg, Fernando. Checkmate to Underdevelopment. Embassy of Chile, Washington, D.C., 1976.
5. Crosson, Pierre R. Institutional obstacles to expansion of world food production. Science 188:4188, 1975.

6. Mellor, John W., et al. Developing Rural India: Plan and Practice. Cornell University Press, Ithaca, 1968.
7. Flaschen, Steward S. and Robert H. Cotton. Foundation Chile, an experiment in technology transfer. ITT Paper, 1977.
8. Rose, Sanford. Why the multinational tide is ebbing. Fortune XCVI:2, 1977.
9. Charron, Ernest C. Financing food production in the LDCs. Chase Manhattan Bank. Paper presented at Conference on Agribusiness 1977 and Beyond, Chicago, April 25-26, 1977.
10. Rose, op. cit.
11. Stobaugh, Robert B. A proposal to facilitate international trade in management and technology. Paper completed under the auspices of New York University Graduate School of Business Administration project: The Multinational Firm in the U.S. and World Economy, 1973.
12. Sametz, Arnold W. The decline of private foreign investment in the LDCs -- Causes and cures of the widening gap. Saloman Brothers, Center for the Study of Financial Institutions. Working Paper No. 2, New York University Graduate School of Business Administration, April 1973.
13. Charron, op. cit.
14. Rose, op. cit.
15. Wall Street Journal, November 10, 1977.
16. Sametz, op. cit.
17. Monckeberg, op. cit.

Part 2

Reducing Climate's Impact

Lloyd E. Slater

7. Three Approaches to Reducing Climate's Impact on Food Supplies

Introduction

What can be done by farmers and governments to make food production resist the impact of undesirable climate and bad weather? In our work in the Food and Climate Forum of the Aspen Institute for Humanistic Studies we have identified three kinds of effort:

1. Techniques which improve climate-marginal food production.

2. Policies and strategies which minimize the impact of future bad weather.

3. Technology which makes food production essentially climate-independent.

Techniques to improve food production in regions of harsh climate or vacillating weather include more efficient irrigation, climate-defensive agronomics, genetic improvements which resist unfavorable weather, and the management of crops with information systems. The use of computer-based climate/agronomic models to optimize agricultural production in climatically difficult regions appears especially promising in the underdeveloped tropics. For this reason the Forum has a project underway involving use of this technology in the llanos, or savannah regions, of western Venezuela -- an experimental area representative of about 300 million hectares of potential new food-producing land in South America. (A brief description of this project in Venezuela appears in Chapter 9.)

Policies and strategies to minimize the impact of future bad weather are those intended to stabilize and extend food supplies, usually at the national level. Possibilities include accumulating national grain reserves, bilateral trade arrangements, integration of ruminants into the food system, reducing post-harvest losses, and the planned, widespread application of food-processing industry technology.

Finally we must consider technologies which produce food essentially independent of climate and associated environmental influences. Prospects here include intensive greenhouse or controlled environment food production, culture of aquatic organisms, biomass conversion to food, and the factory manufacture of synthetic food components. While rising energy costs and loss of prime farmland are slowly constraining conventional agriculture, they are also nudging many of these climate-independent food technologies closer towards cost-effectiveness and commercial reality. Carl N. Hodges and the author both discuss some of the more promising options in the final two chapters in this volume.

Improving Climate-Marginal Food Production

More Efficient Irrigation

Irrigation, or artificially supplied water, is widely used to minimize the impact of excessively dry climate or unusually dry weather on food crop production. Approximately 17 percent of the world's agricultural lands, about 250 million hectares, are now under irrigation. Herbert Riehl estimates that this small fraction of cultivated land under irrigation accounts for one-half the world's harvested crops (1). So the potential for adding to world food production by more irrigation is enormous, particularly since there are as yet millions of nonutilized arid and semi-arid hectares which could grow crops if irrigated.

Unfortunately, most of the world's easily available fresh water is already taken up in irrigation or, increasingly, by human settlements and industry. Although there is a vast amount of untapped water underground, and many rivers still pour virtually unused to the sea, the cost of raising or transporting it to grow crops is in most cases prohibitive.

There is, however, great hope for expanding world food production by improving the efficiency of irrigation as it is now practiced. Victor Kovda, the eminent Soviet soil scientist, estimates that less than 50 percent of all water deployed in irrigation is being used efficiently (2). Further, he notes that over 60 percent of all irrigated lands are gradually losing their fertility because of waterlogging and salinity. Irrigation mismanagement and poor technology now cause a loss of over 200 thousand hectares of good farmland each year (an area about the size of Luxembourg) (3).

Figure 1 shows the number of irrigated hectares as a percentage of all land cultivated in eight important farming nations. As per FAO figures in 1971 it should be noted that an estimated 25 million new hectares of irrigated land were added in the 1970s, many in the countries indicated. Not shown in Fig. 1, however, is the relative efficiency in the way these countries manage their irrigation systems. China, which now irrigates over 69 percent of its agricultural land, apparently meets almost all food needs for its over 900 million people from about one-half the land area now under cultivation in the United States. In China, irrigation is highly efficient, with good drainage a long-established technology. Egypt, less successful in feeding its 40 million people without large food imports, irrigates all its cultivated land. In Egypt, as a result of its High Aswan Dam project, water is free to anyone with power to lift it to the fields; hence, with animal power plentiful, overwatering of crops is widespread. Besides water waste, poor drainage and subsequent soil salinization have steadily decreased fertility in what was once the richest food-producing area in the world. But there are promising signs that Egypt is mastering its irrigation problems in a widespread program of on-farm management (4).

Techniques for improving the efficiency of existing irrigation systems are well described in an excellent booklet by the National Academy of Sciences, "More Water for Arid Lands" (5). At least a dozen well-known ways to reduce water losses are covered. Drip irrigation, first used on a large scale in Israel, reduces waterlogging and saline build-up, while reducing the amount of energy and water required by as much as 50 percent (6). Today's costs for improving or modifying existing water systems are high but affordable, given the almost immediate return in water savings and crop productivity.

Climate-Defensive Agronomics

Field management techniques, such as those revealed in Fig. 2 (an intercropped field at the International Rice Research Institute in the Philippines), are an effective way to reduce the risk of crop failures due to unfavorable weather. Research on mixed cropping, traditionally used in the tropics but rarely practiced in large-scale temperate monoculture farming systems, proves that food crops do better with fewer inputs and are more resistant to climate stress when interplanted on the same plot of land (7). A selected mix of

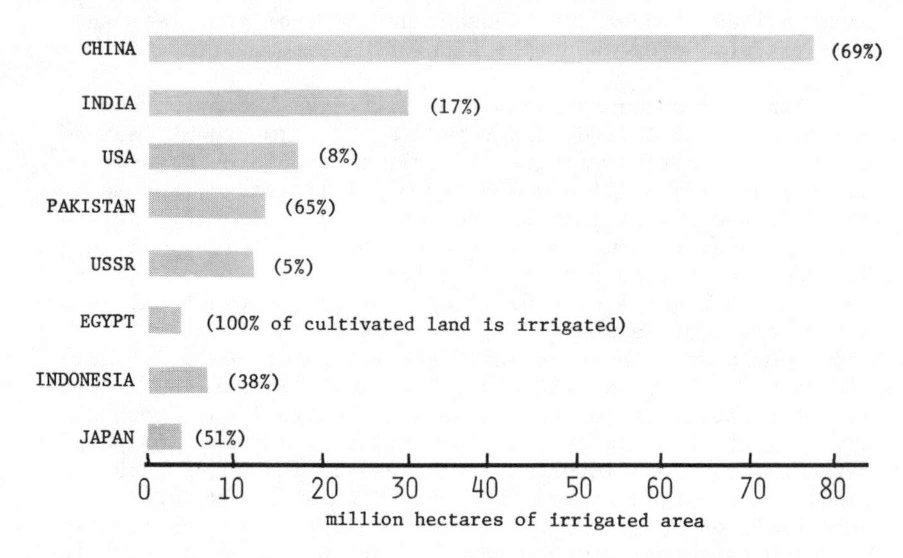

Fig. 1. The number of irrigated acres as a percentage of all land cultivated in eight important farming nations. Source: FAO, 1971.

Fig. 2. An intercropped field at the International Rice Research Institute in the Philippines demonstrates an effective way to reduce risk of crop failure due to unfavorable weather.

plants with varied foliage and root systems creates its own optimum microclimate, reducing excess solar exposure and water needs, curbing weed and pest problems and sometimes increasing individual plant yields. Its root system efficiently utilizes soil volumes, minimizing erosion during intensive rainfalls. Further, while unexpected climatic stress may drastically reduce yields and even wipe out one crop in the mix, the farmer will usually get a useful harvest from what remains.

Agricultural scientists have been studying the relationships between crops and their environment to find ways to reduce water needs without decreasing photosynthetic efficiency. Experiments with windbreaks, for example, have shown that devoting a small portion of the farming area to this barrier will alter the microclimate over a field enough to lower temperature in foliage and reduce water vapor transpiration by as much as 20 to 30 percent. Stubble mulch farming and minimum tillage are other long-known practices, now being revived, which significantly improve soil moisture conservation. Norman J. Rosenberg reviews these and other research strategies in his chapter later in this volume.

Recent work by the Center for the Biology of Natural Systems at Washington University disclosed an unexpected advantage of organic farming systems over equivalent-sized energy-intensive conventional farms in the United States cornbelt. Foregoing synthetic fertilizers and pesticides and rotating their crop fields to legumes and hays to conserve soil nitrogen, the organic-type farmers, while they received less net income compared to conventional farmers during good weather years, were found to average 4.5 percent higher net return in income during bad years when weather was a dominant factor. Improved soil moisture conservation and drastically reduced erosion losses were credited for this climatic advantage (8).

Genetics to Resist Climate

While the main thrust in food production genetic research has been to increase yield, there is much promise for breeding plants and animals to improve their resistance to the temperature and precipitation extremes of unfavorable weather and climate. Fig. 3, for example, shows the improvement in rice yields in Colombia following introduction, in the late '60s, of dwarf varieties which resist lodging caused by heavy rains and winds. Of course, the primary intent in this genetic improvement was to increase yields in a heavier head of grain; the shorter stem was necessary to carry this new load.

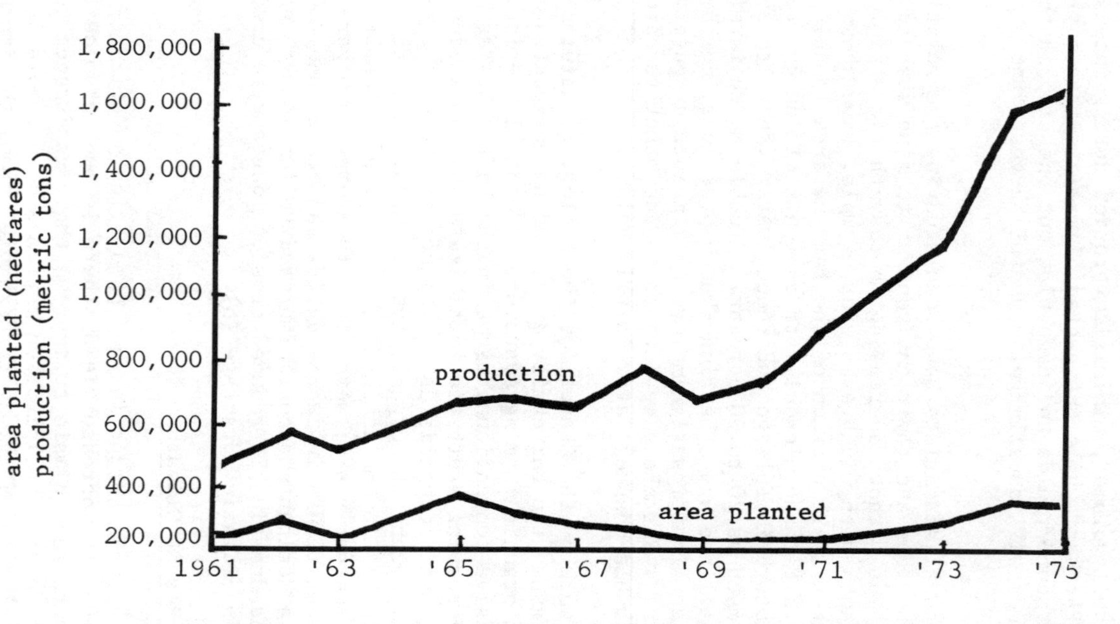

Fig. 3. Improvements in rice yield in Columbia following introduction of dwarf varieties which resist lodging caused by heavy rains and winds. Source: Scientific American, 1976.

Worldwide use of new high-yielding hybrids has resulted in a dramatic success story. Between 1965 and 1975 acreage in HYVs of wheat and rice (excluding communist nations) increased from 59 thousand to 41 million hectares, with yields often more than doubling (9). However, greater susceptibility to drought and pests, as well as increasing input requirements in many of the new hybrids, has caused a point of diminishing returns. Research efforts directed towards improved climate-response factors, both in crop and livestock hybridization, may help reverse this trend.

For example, work at the University of Nebraska has aimed at modifying a plant's ability to reflect solar radiation in order to reduce transpiration, while not affecting the rate of photosynthesis (10). Experiments have shown that natural clay coatings on some plants reflect light, hence effectively reducing the amount of water required in maintaining plant temperature. While the technique of massive clay application is undoubtedly prohibitive, the principle itself may be used in developing water-conserving hybrid strains.

In a provocative recent paper on the consequences to agriculture of increasing carbon dioxide in the atmosphere (11), Sylvan H. Wittwer suggested that a promising research initiative could be mounted using genetic resources to alleviate climatic stress. He pointed out that inevitable increasing levels of CO_2 in the atmosphere will contribute significantly to photosynthetic performance, and that genetic responses to high levels of this gas should be studied in terms of water-use efficiencies. Wittwer concluded that climate change should not be viewed exclusively in a negative context and implied that the science of genetics could be exploited to achieve favorable agricultural response.

Still another promising path for using genetics to deal with unfavorable climate in food production has turned up in intensive new studies of natural halophytes, or salt-tolerant plant species which have potential value as terrestrial food crops. Work in Israel, Delaware, California and in Mexico suggests that selection and further breeding of these species will yield commercial production of grains, seed-crops and tubers on vast areas of arid and desert lands with seawater irrigation (12). In Chapter 12, Carl Hodges reviews some of the possibilities.

Agro-Climatic Information Systems

Computer-based agro-climatic information systems are a relatively new and promising technology for reducing the

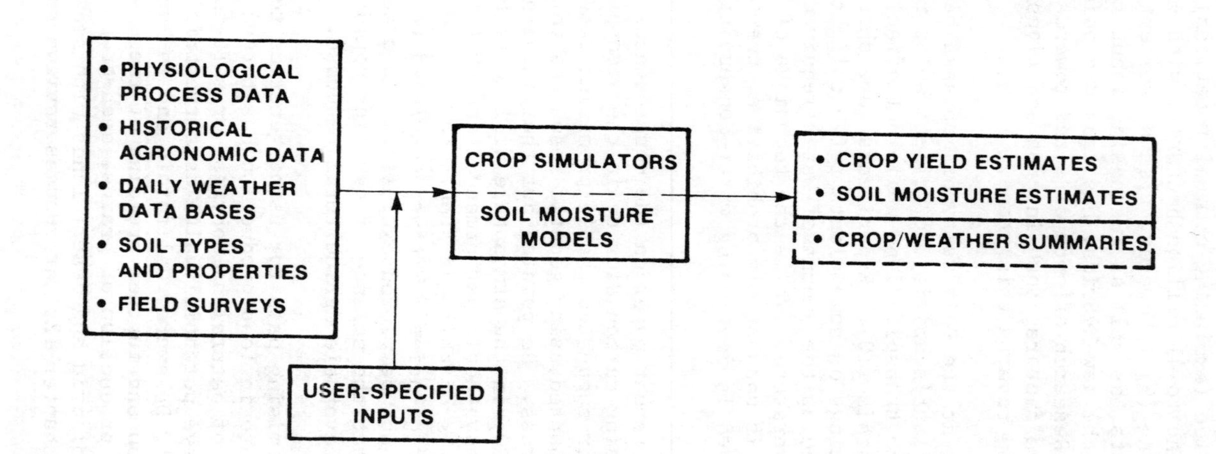

Fig. 4. Elements in a computer-based agro-climatic information system, in this case Control Data Corporation's AGSERV, which simulates corn and soybean production in the U.S. Midwest. Courtesy Control Data Corporation.

vulnerability of food crops and food systems to unfavorable climatic conditions and unexpected runs of bad weather. In the chapter which follows, James D. McQuigg describes some recent applications.

There has been a wide variety of efforts in integrating climatic information and weather models with agricultural data for simulation, prediction and analysis purposes. A recent compilation by the United States Department of Agriculture lists 183 research and experimental projects in this area during the period 1970 to 1978 (13). In its most practical approach, the computer holds a model of a farming region's past climate, agronomic resources and crop characteristics. It supplements this with real-time information on prevailing weather, soil conditions and other factors that determine strategic times for inputs such as fertilizer or irrigation. The system advises when to plant, harvest and/or replant in the face of impending drought or flooding conditions, and even suggests what crops, varieties and species to plant for maximum performance in a climatically difficult region.

Fig. 4 illustrates the elements in a typical agro-climatic information system: Control Data Corporation's AGSERV, which has been under test and development for the past three years. AGSERV incorporates information from over 1,000 counties in the United States cornbelt. While the computer, with its physiology-based crop models and data bank, is located in Control Data's Minneapolis headquarters, subscribers have access through their own remote terminals.

As Dr. McQuigg indicates in the next chapter, Aspen Institute's Food and Climate Forum at present has a project underway in Venezuela to determine the value of agro-climatic information systems technology in a relatively undeveloped farming region in the tropics. While the project is not an attempt by Aspen Institute to "transplant" an AGSERV-type system as such, Control Data has generously offered the expertise behind this system and computing services in Venezuela to the project. It has become clear, thus far, that such technology, developed for temperate-zone conditions and advanced country infrastructure, is not portable; that in order to be successful the information system must be designed and assembled almost from scratch, and its programs adapted to the special conditions and needs of the tropics. Thus far, as the project proceeds, there is every reason to believe that the resulting system will not only perform well, but also furnish returns of even greater value than in the already well-organized and highly institutionalized setting of temperate-zone agriculture.

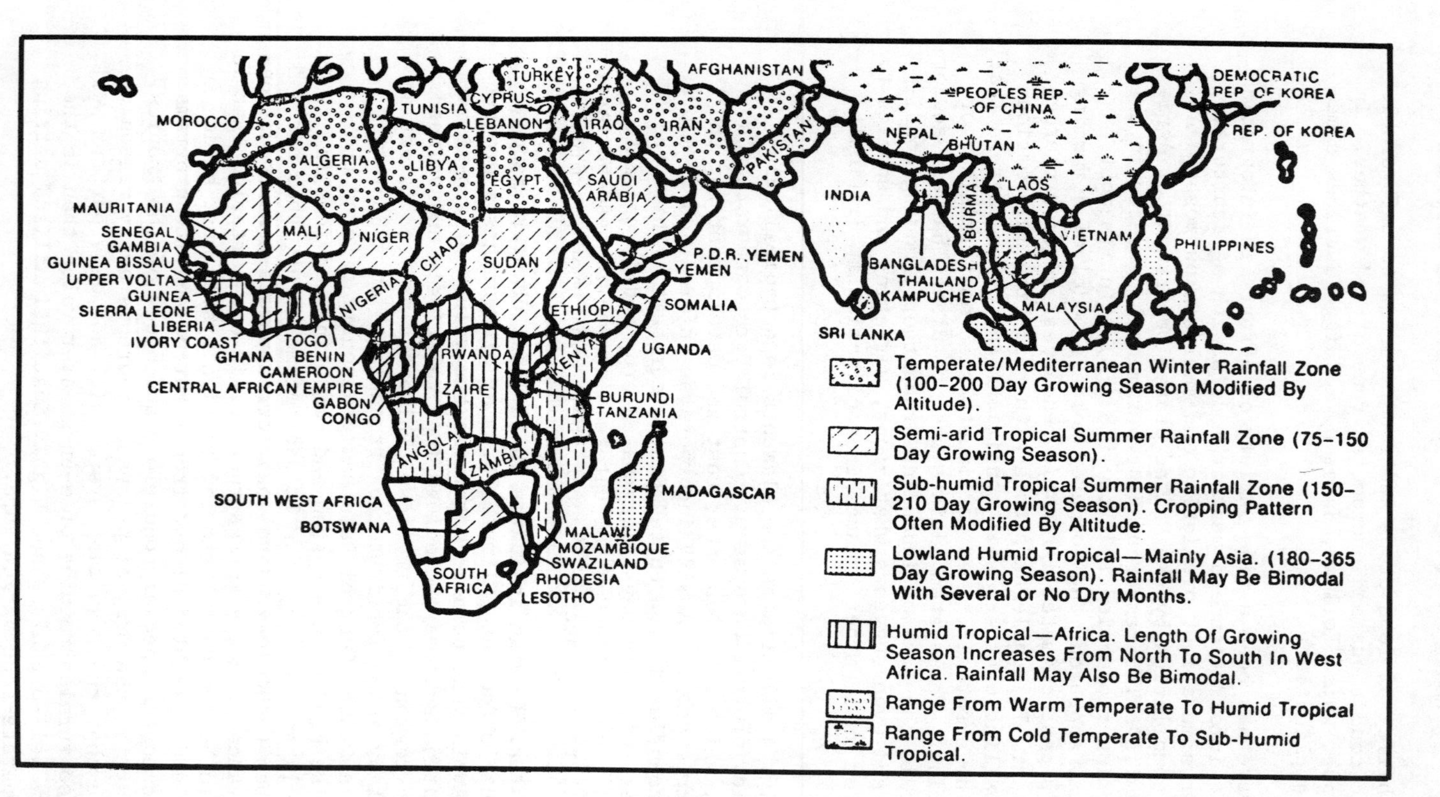

Fig. 5. As the map discloses, at least five different climate regimes prevail in the developing countries of Africa and Asia. Far too often the dominant crops grown are inappropriate under local climatic conditions. Source: International Food Policy Research Institute, 1975.

More Appropriate Land Use

A final example of climate-defensiveness in agriculture, often overlooked because it is so obvious, is the concept of environmentally appropriate use of the land. Fig. 5 shows climate regimes in the developing and usually food-needy countries of Africa and Asia. Far too often what is grown and the use of land are inappropriate under local climatic conditions and unmatched to local food needs.

Land use priorities in Mali, West Africa are an example. In that predominantly dry country, cotton and rice, which have high water requirements, are extensively grown. Yet during the severe Sahelian drought, lasting from 1968 to 1973, Mali cotton, rice and groundnut production, all for export, was not allowed to suffer for lack of water and actually reached record highs. But corn production, the predominanat food staple for most, fell by one-third and many starved (14).

The pattern of climatically inappropriate cash crops dominating the best lands at the expense of indigenous food production still prevails throughout most of the tropical developing world. It is a carry-over from the colonial system, where commodity-type monoculture was introduced to generate foreign sales and, since independence, foreign exchange. As a result, most of the research effort and infrastructure in those countries favors crops such as coffee, palm oil, rubber and cotton rather than locally-consumed foodstuffs. In Tanzania, for example, during the crop year 1973-'74, just before its massive importation of emergency food supplies, the country exported well over 700 thousand tons of cotton, coffee, sisal and cashew nuts (15).

There are a large variety of food crops which can be grown for local consumption and even for export which more properly meet the climate and environmental constraints than the crops which now prevail in the climate regions shown in Fig. 5. In the arid Sahel, for example, any of the millets, which are the most drought and heat resistant of all cereals, could be extensively cultivated. That this can be done as a national effort was well demonstrated in Senegal, where its Institute of Food Technology at Dakar (ITA) first developed a millet bread and gained its acceptance with consumers before the government embarked on a widespread program to encourage millet farming (16).

Hedging Against Future Bad Weather

As Fig. 6 reveals, since World War II, mainly due to

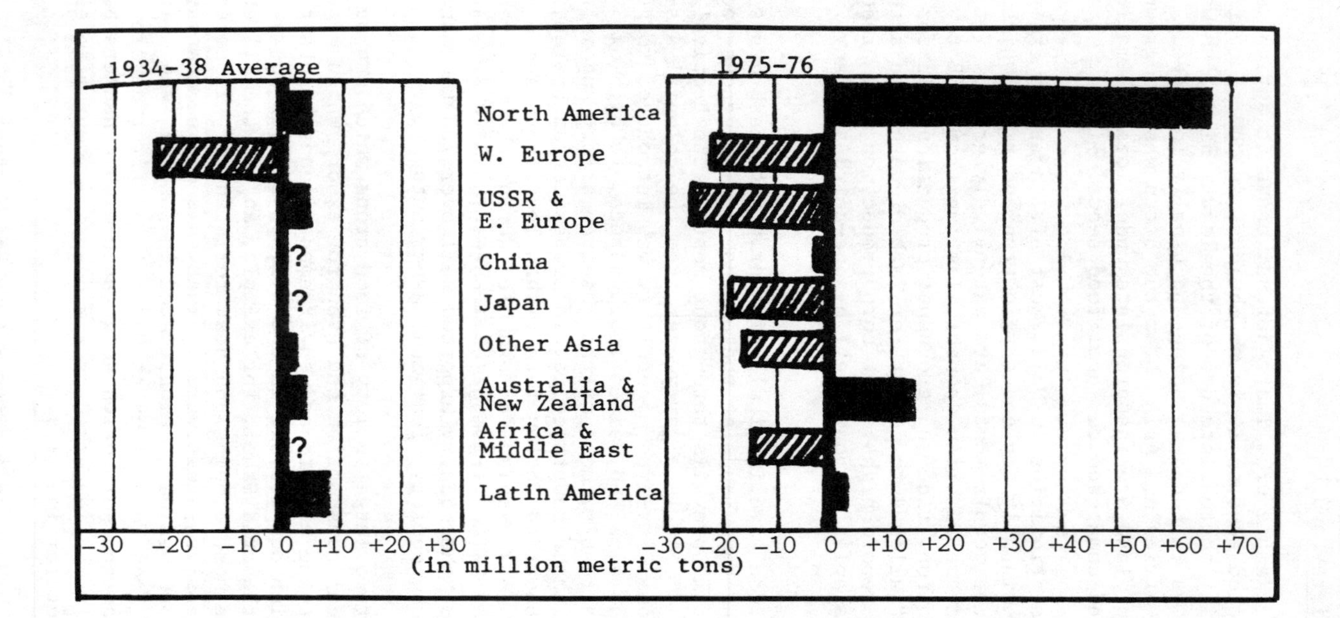

Fig. 6. Changes in World Net Grain Trade, 1934-1976. Since World War II most nations of the world have come to rely on imported grain from a few major producing countries. Adapted from Scientific American, September, 1976.

increased population demand, but often poor agricultural planning, almost all nations have come to rely on imported grain from a few major producing countries to supplement their own food production. In Chapter 1 in this volume, F. Kenneth Hare describes how this shift away from local food self-sufficiency, coupled with the impact of normally reoccurring bad crop-growing weather, gravely threatens food security, particularly in the poorer, tropical countries. A sequence of bad weather in major grain-growing regions, such as the USSR, China, or U.S.A., causing a worldwide production dip of only one or two percent, escalates import prices beyond the means of the poorer countries. When this coincides with local food crop shortfalls due to bad weather, vast numbers in the population suffer increased undernourishment or outright starvation.

Accumulating Food Reserves

As Martin Abel points out in Chapter 5, there are several measures that can be taken by food-marginal nations to hedge against the impact of unsupportive future climate and runs of bad crop-growing weather. Most obvious, although not necessarily easiest to implement, would be a national effort to establish continuing food reserves through policies which favor overproduction and storage of surplus in good crop years to compensate for shortfalls when the inevitable bad run of weather occurs.

India, with its unpredictable monsoons and their powerful influence on crop success, started its effort to accumulate and properly store foodgrain in the early 1970s. Fig. 7 shows the results. In 1978, boosted by three years of bountiful harvests due to well-timed and adequate monsoons, reserve stocks reached 20 million metric tons. These reserves sustained India following the growing season of 1979-1980 when monsoon rain proved inadequate.

A major reason for the sharp gains in India's surplus production, despite its steadily rising population, was the new agricultural policy instigated by the Janarta government in 1977. It established incentives to increase local production, while at the same time building up reserves which would carry the nation through two years of inadequate monsoons. The government estimates it needs at least 12 million tons of foodgrains in storage to achieve this end. Excess amounts of reserves are considered unwise because of post-harvest spoilage and the immobilizing of large amounts of money and credit. In India the new national policy has helped to ease factional tensions in the population, which now takes great pride in no longer being dependent on imported foodgrains.

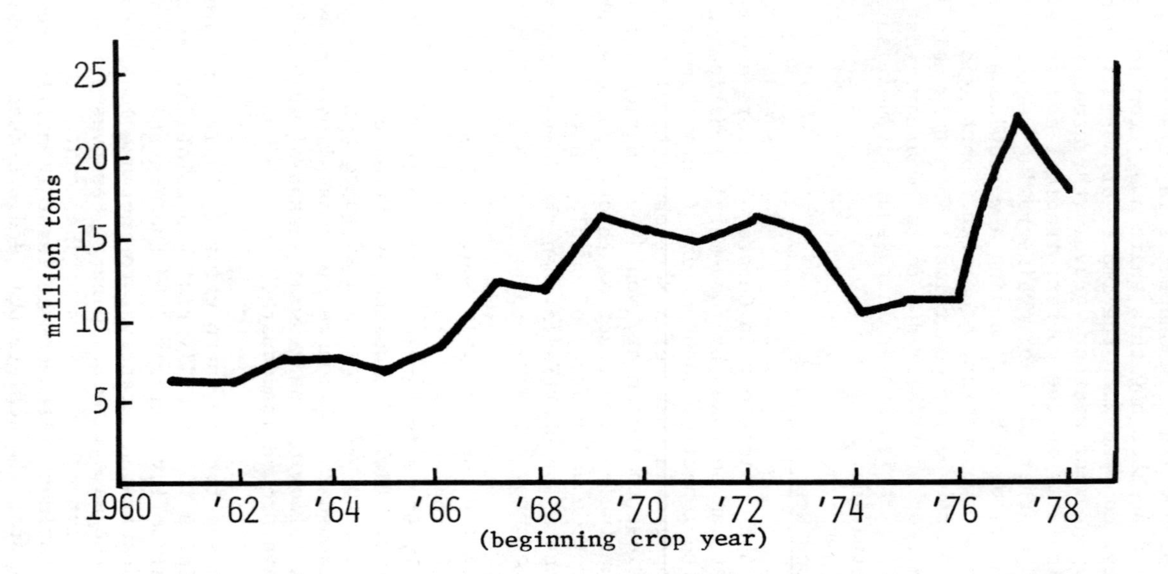

Fig. 7. Foodgrain Stocks in India. India's program to establish enough grain reserves to carry the nation through bad monsoon seasons, established in the mid-1970s, paid off during the poor growing season of 1979-80. Based on American Universities Field Staff, 1979.

It has also helped to stabilize and hold down prices in times of weather-induced shortfalls, thus improving purchasing power among poor countries.

Many nations, particularly the smaller ones with limited arable land resources, lack the capacity to accumulate food reserves through their own production. Often, when demand elsewhere has preempted surplus stocks or driven prices beyond reach, the poorer, developing nation faces a famine situation and must rely on emergency food transfers to meet its needs.

As Martin Abel indicates in Chapter 5, a policy of bilateral trade offers the small agricultural nation a way to avoid periodic food shortfall crises. The arrangement shown in Table 1, intended for use between the United States and Bangladesh, suggests a model for similar agreements between export and import nations. It offers a flexible mix of purchasing schemes, including straight commercial, U.S. Public Law 480 (or liberal credit), and concessional, depending on the world market price.

In bilateral agreements, maximum and minimum purchases are mutually established. This protects the exporting nation by guaranteeing a relatively steady market. The minimum level requirement serves the import nation in the long run. During bountiful crop years it is forced to build up its own foodgrain reserve, thus enhancing its ability to meet unusually heavy shortfalls in the future.

Reducing Post-Harvest Losses

The success of food reserve policies greatly depends on adequate and secure facilities to store the locally-grown or imported grain. Without such facilities post-harvest losses can often be large enough to negate benefits of a food-reserve program.

Table 2 shows the weight losses after harvest for various crops in several different food-growing regions of the world. As the table reveals, such losses may run as high as 50 percent or more in countries in the humid tropics where the poor food containment and warm damp climate encourage fungi, insects and hungry rodents. One should note that such losses are almost minimal in temperate zone, developed nations, with their cooler climates and good storage facilities.

To comprehend the massive extent of food waste, one should consider the conservative estimate of 25 percent

Table 1. A bilateral agreement, such as this suggested one between Bangladesh and the U.S., offers a flexible means for assuring food security despite climatic impact on supplies at both national and international levels. Source: Schnittker Associates, 1978.

World Wheat Price Level ($/bu.)	Sales Terms Under a Bilateral Agreement with a Low-Income Developing Country		
	Commercial	P.L. 480	Donation
2.25-3.15	15 %	60 %	25 %
3.15-3.50	10	63	27
3.50-3.85	5	66	29
3.85-4.20	0	70	30

Table 2. Weight losses in foodcrops after harvest are especially high in tropical developing countries where storage facilities are minimal and climatic conditions favor biodegradation and foraging by fungi, insects and rodents. Source: FAO, 1970.

Crop	Country	Weight Loss (%)	Storage (# of months)
Legumes	Upper Volta	50 - 100	12
	Tanzania	50	12
	Ghana	9.3	12
Maize	Zambia	90 - 100	12
	Benin	30 - 50	5
	USA	0.5	12
Rice	Malaysia	17	8-9
	Japan	5	12
	United Arab Republic	0.5	12
Sorghum (unthreshed)	Nigeria	2 - 62	14
(threshed)	USA	3.4	12
Wheat	Nigeria	34	24
	USA	3.0	12
	India	8.3	12

post-harvest loss in cereal and pulse crops in semi-arid Africa. Based on 1971 production figures, such losses amounted to about 7 million tons. At an average of $140/ton, this amounts to about a $1 billion loss in food in just one poor region of the world. Since the calories and protein available in one ton of cereal will furnish most nutritional requirements for six people for one year, the food saved could supply most of the food needs of millions (17).

Most developing countries lack even rudimentary facilities for drying, preserving and properly holding food after harvest. Grains and root crops brought in from the field are usually piled in flimsy sheds or lean-tos until needed, or stacked directly exposed to weather. Steady attack by insects and rodents on the food stockpile is an accepted fact of life. But a continuous loss of the food's quality and nutritive value also occurs at the same time. Overexposure to the sun and heat destroys vitamins and, coupled with humidity and dampness, produces undesirable chemical and texture changes in the food. Fungus and rot blight the surface and insides of stored food, and rodent droppings and insect parts are surrounding contaminants.

The need for simple, effective on-farm and village level storage structures is a worldwide imperative, since about 80 percent of the average farmer's grain crop is retained at these levels. The technology needed for rural storage facilities has been available for some time, although it has found only token use in most regions of the world. Still, the success when used has been dramatic. For example, the agricultural extension service in Benin, Nigeria, assisted by volunteer workers and aid agencies, has carried out a low-cost food-storage building program that has almost eliminated rodent foraging in that area. In northern Ghana a private technical assistance group working with a church agency has adapted the traditional mud silo into an effective long-term holding facility. Both in Nigeria and Ghana, the levels of engineering and construction for the storage units were simple enough to be taught to local farmers, with materials used mainly from local and inexpensive sources.

While loss of food due to environmental degradation and hungry predators is generally heaviest during its post-harvest period of storage, sizable additional losses can also occur anywhere along the food delivery chain, stretching from the rural field to the remote urban consumer. Table 3, indicating such losses in the handling of rice, shows how well over 10 percent of Southeast Asia's crop disappears before it reaches the consumer.

Table 3. Total losses in rice in Southeast Asia from handling and processing after harvest are estimated as high as 37 percent by researchers at the University of the Philippines.

Operation	Range of Losses (%)
Harvesting	1 - 3
Handling	2 - 7
Threshing	2 - 6
Drying	1 - 5
Storing	2 - 6
Milling	2 - 10
Total	10 - 37

Source: International Development Research Center, 1976.

<u>Completing the Food System</u>

Most developing countries, particularly those in the tropics, have primitive food systems, with production for internal consumption mainly in the form of raw, unprocessed products. Facilities for storing, preserving, packaging and distributing food, and their supporting infrastructure are largely missing. Hence, native food supplies come in seasonal bursts, depend on marginal and fragile distribution paths, and are especially vulnerable to losses due to bad weather and unfavorable climate.

Fig. 8, diagramming a fully-developed food system in a modern, industrialized nation, shows the steps beyond the farm gate and related alternative food-growing subsystems that are usually lacking in lesser-developed countries. Creating such a system -- one which closely integrates what is grown and processed with national needs and constraints -- offers many interesting new options in climate-defensiveness and food self-sufficiency to the developing nation. For example, the systems approach and dedicated introduction of food-processing technology opens the way for new and more appropriate choices in what is grown at the farming end of the system. At present many countries in the tropics devote their best growing land and supportive policies to the non-food commodity crops produced for export. A systems analysis of national food capabilities and needs may shift priorities away from export plantation farming towards the organized crop culture required to supply food-processing plants and a sizable domestic market.

Food processing -- its preparation and packaging in preserved form -- not only adds value and convenience to the product, but also extends its life for several years of safe storage. Furthermore, the investment in processing technology, which for efficiency should be near to the raw food sources, encourages rural development and requires supporting infrastructure and distribution/marketing facilities which stabilize supplies in urban areas. Farmers, assured of income from processing plants for their cash crops, can secure the credit required for agricultural inputs which can be used to minimize the impact of bad weather or a difficult climate.

As Fig. 8 suggests, there are many prospects for using modern food process engineering in imaginative and creative combinations as the crops of poor countries are industrialized. Processes and products needing high energy or capital input will undoubtedly prove too costly. Newer technologies such as irradiation, aseptic packaging and bio-synthesis may be found, in cost benefit analysis, to have special

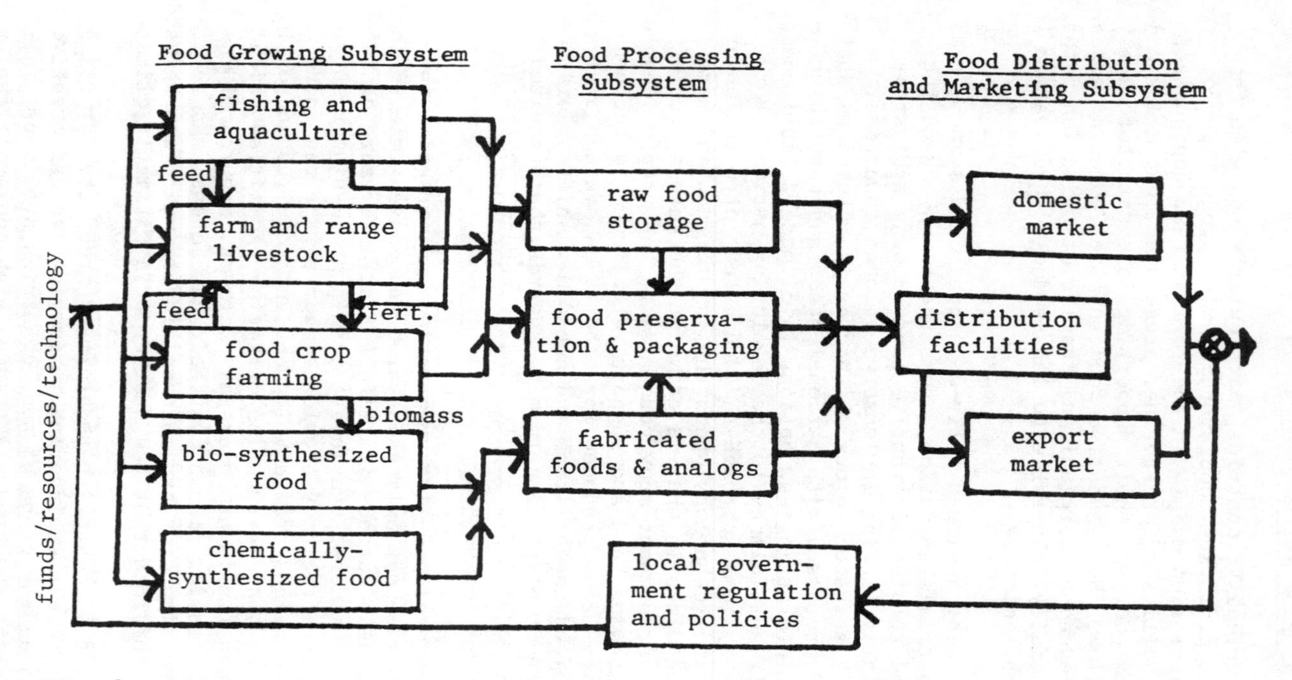

Fig. 8. Diagram of a fully developed food system in a modern industrialized nation shows the steps beyond the farm gate and related alternative food-growing subsystems that are usually lacking in lesser developed countries. Source: L.E. Slater, 1978.

justification in the mainly tropical, developing world. Extensive refrigeration and elaborate packaging, so well established in affluent nations, may be bypassed. Bulk-containerized sterile foods along with innovative low-cost distribution techniques could prove a viable combination. Single-cell protein, grown by fermentation on the prolific tropical biomass, could become a significant local food source and even a rewarding export crop in many developing countries (18).

The Special Role of Livestock

Many futurists, viewing with alarm the food production vs. population equation, suggest that livestock raising and meat will gradually disappear in the world food system; the grain now fed to animals must inevitably go into the human diet. What these visionaries ignore, however, is the significant role that livestock, especially ruminants, could play in the necessarily low-energy agricultural systems which many also deem are inevitable in the future.

Fig. 9 shows how ruminant livestock occupy a central place in a well-integrated food-producing system. Besides providing meat, milk, fertilizer and other useful byproducts, they serve as capital reserve in lean years. During crop seasons when good weather favors surplus harvests, livestock herds are fed on the surplus and increase. When crops fail and there is not enough food for local people, more meat becomes an important factor in the diet and herd sizes dwindle.

Because of their compartmental stomachs, ruminants can sustain growth on grasses and other wild forage in areas too dry and otherwise unfit for cultivation. Ruminants such as zebu cattle, goats, reindeer and camel are well adapted to special climate regimes and can survive where other livestock fail. Twelve percent of the world's population get their support almost entirely from ruminants because they live in areas where food crops cannot be grown. The fat stored in ruminants, most obviously in the hump of the zebu or camel, supplies energy when feed is scarce. During periods of drought in certain countries in Africa, for example, the goat has been one of the few animals to survive and has, in fact, been charged with aggravating the process of desertification. Yet the goat is the only animal which can subsist on such meager resources, and its presence has often meant the difference between subsistence and starvation (19).

A livestock/crop rotation system serves many uses. It contributes to soil fertility and erosion control and helps

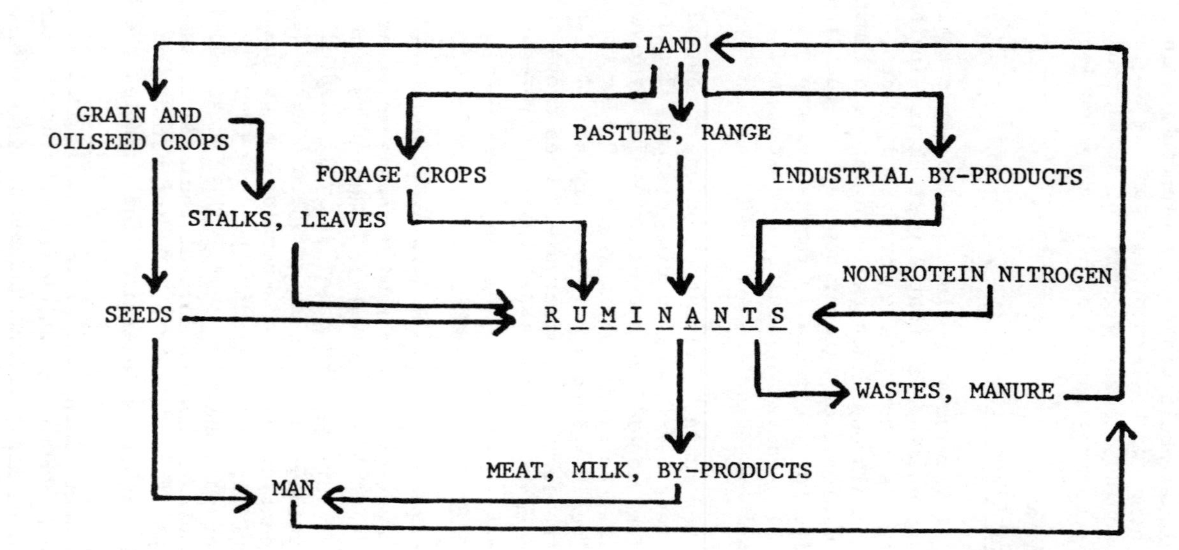

Fig. 9. As the diagram reveals, ruminant livestock play a central role in a well-integrated food system as well as serving as a buffer during years when crop yields suffer due to bad weather. Source: Agricultural Council for Science and Technology, 1975.

check soil-borne diseases of both crops and animals. When rangeland is properly managed, grazing serves as a stimulus to productivity of the herbage. Furthermore, ruminant diets can be exceedingly flexible. Although in the developed nations they receive about 30 percent of their feed in grain, they can be, and are increasingly, fed a wide range of agri-industrial byproducts which often go unused in many developing nations.

Producing Food Independent of Climate

The climate factor in food production, like a fierce mother-in-law in the household, can be dealt with in two ways. Either you learn to live with it, or you try to get rid of it.

Getting rid of the climate factor, however, is not all that easy. Like the dedicated mother-in-law, it has a pervasive influence which somehow extends into the immediate growing environment, no matter how well quarantined. Take greenhouse culture, a long-established technique for eliminating the vagaries of unfriendly climate. Yet to be successful the greenhouse must still have a dependable supply of free solar energy if climate allows. Even culturing fish beneath a protective canopy of water doesn't completely eliminate the problem, as the climate-induced vicissitudes of El Niño and its impact on anchovita production so well demonstrate.

Producing food inside completely closed containers, such as huge fermenting vats or chemical reactors, would seem to be as climate-free as one could get. Yet both technologies rely on raw materials or substrates which must inevitably come from sun-soil-and-rainfall agriculture. So their feasibility and success are still conditioned by how well climate cooperates somewhere on this planet.

Nevertheless, with world food needs ponderously growing despite limits in land, water and energy, and with no sign that climate will suddenly become uniformly benign, one must seriously consider unconventional ways of supplying an increasing share of food in our future.

Controlled Environment Agriculture

As Fig. 10 suggests, the conventional greenhouse, where plants are grown on soil with temperature and moisture maintained at near-optimal levels, is moving these days towards systems of completely Controlled Environment Agriculture (CEA). The system illustrated is one recently developed by

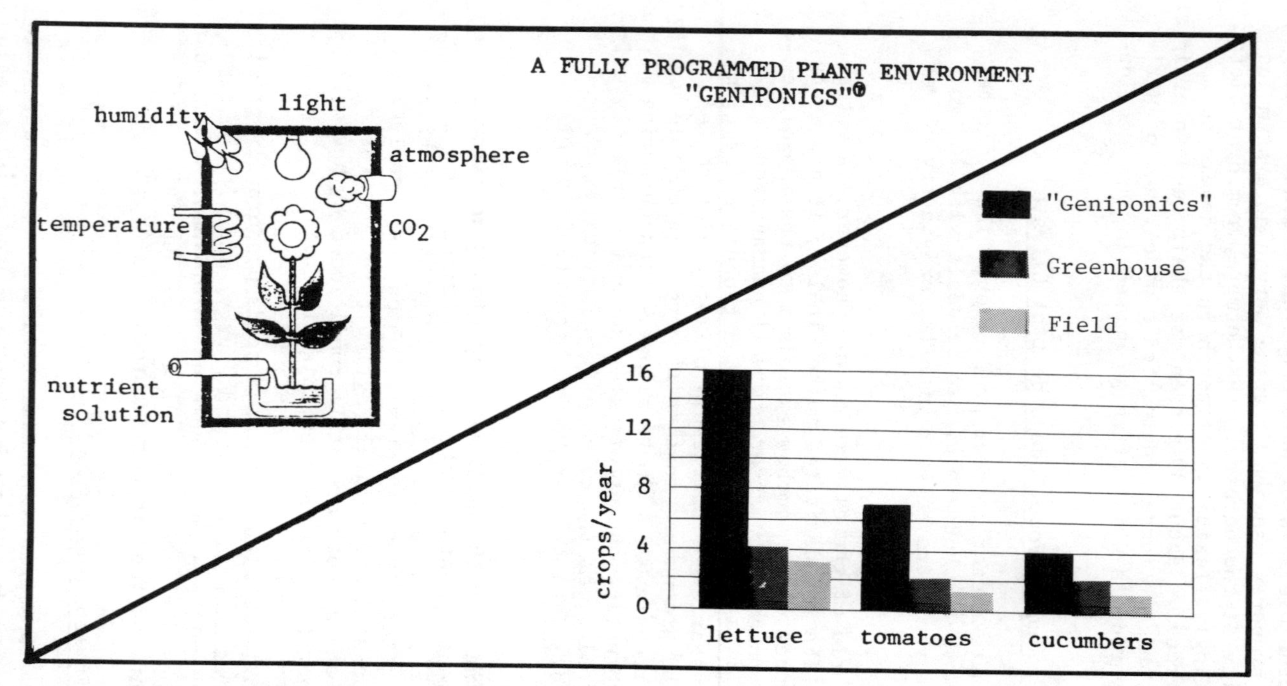

Fig. 10. An example of controlled environment agriculture recently developed by the General Electric Company is this system which controls all environmental factors, including light emission. Major improvements in yields over conventional greenhouse culture are claimed. Source: General Electric Company, 1978.

the General Electric Company as a potential new market for its illumination technology. Utilizing programmed high-intensity lighting along with controlled microclimate, nutrients and water yields, says G.E., results in three or more times the productivity possible in equivalent greenhouse culture. The approach appears to have great promise in affluent cold climate locations where energy is in good supply and there is avid demand for fresh produce. In 1978 General Electric started growing tomatoes by this method in a 16,000 square foot pilot production facility in Syracuse, New York. It has sold one installation to the U.S. Army, now experimentally growing vegetables for troops stationed in Argentia, Newfoundland.

Another interesting entry in CEA was introduced by the General Mills Company in 1979, after five years of costly research and development: a highly automated "vegetable factory" located in a $4 million new plant near Chicago, Illinois. The GM system grows each green vegetable in its own nutrient container on a continuous-belt operation over 30 days. The initial crops, which General Mills believes will compete favorably in the quality urban salad-greens market, include lettuce and spinach (20).

Aside from these rather unique and advanced new systems, controlled environment agriculture in more conventional solar heated facilities has advanced significantly in recent years. Better greenhouse design, coupled with more efficient collection and use of solar energy and new techniques in hydroponic (soil-less) culture, is extending the range of CEA-grown crops successfully marketed. While a sophisticated CEA installation is high in capital costs -- now $200,000 and over per acre according to its complexity -- it is expected that technological advances, economies of scale, conservation of water and nutrients, and alternative uses of the land will make its widespread use feasible and competitive during the next century. As Carl Hodges points out in Chapter 12, CEA greenhouses have special virtue for location on relatively inexpensive arid land, or in desert areas relatively near good markets, where sunshine is abundant and water conservation a necessity.

Aquatic Farming Systems

Shown in Fig. 11 is the conceptual design for an aquaculture system suggested by scientists at Oak Ridge National Laboratory as a way to make use of the "thermal pollution" Btu's in nuclear power plant cooling water. While specific in its application, the scheme hints at the wide range of possibilities for culturing aquatic organisms for food

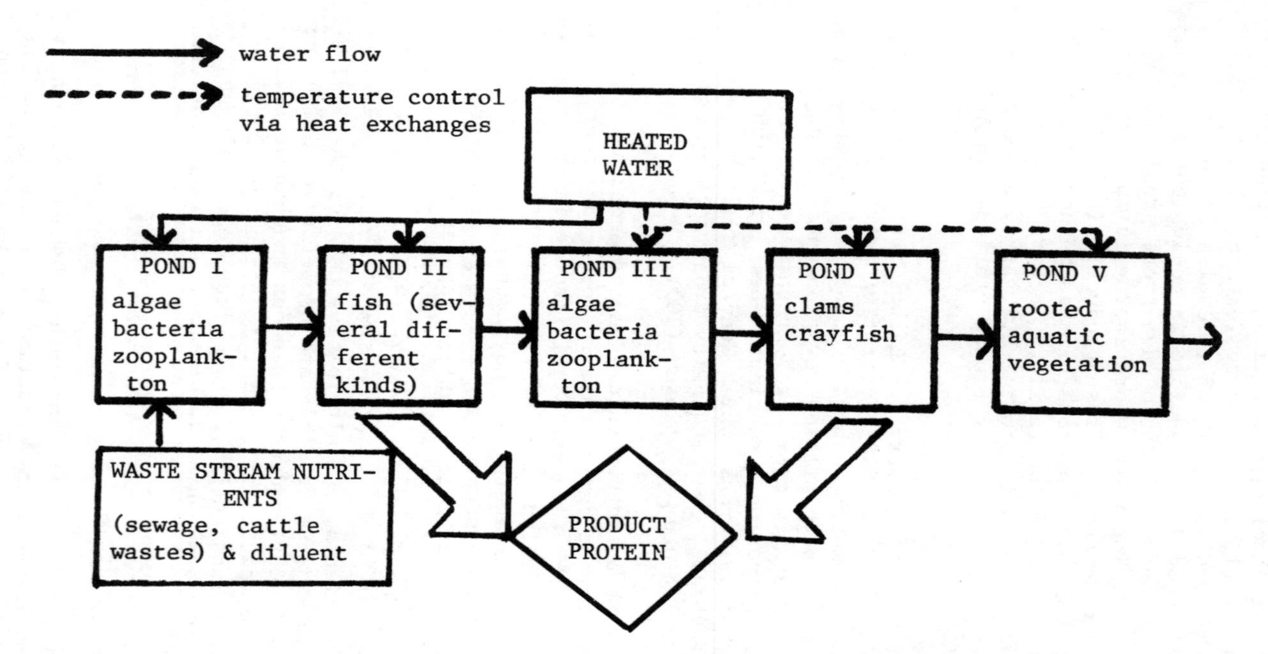

Fig. 11. Design for an Aquatic System. The multiple possibilities of aquatic system food culture are suggested in this scheme developed by Oak Ridge National Laboratory to utilize waste heat from nuclear power plant cooling water. Source: Oak Ridge National Laboratory Review, 1978.

protein using societal waste as a nutrient and energy source. A number of less complex experiments of this type have been underway in the United States during the past few years. Typical is a project conducted by the Texas Agricultural Experiment Station of Texas A&M University at the Cedar Bayou plant of Houston Lighting and Power, where a variety of finfish, crab and shrimp are being grown in the plant spent cooling water outflow (21).

Fish culture, mainly in ponds, hence subject to climate impact to some extent, yielded six million metric tons worldwide in 1975. Recent advances in closed-cycle aquaculture (essentially self-contained fish-growing systems) and in polyculture (raising satellite crops in the system such as disclosed in the Fig. 14 scheme) require significant capital investments and suggest that major food-processing companies must be involved in order to spur the project growth in this source of food to 50 million metric tons by the year 2000, as estimated by the National Research Council in 1978 (22).

The recent escalation in world protein costs, and decrease in conventional fisheries' yields, has fostered this necessary large-scale interest in the productive potential of aquatic farming systems. While a land farmer is lucky to raise more than 100 pounds of beef on an acre of good pasture, a closed-system fish farmer can raise dozens and even hundreds of tons of fish and shellfish on an aquatic acre in a year. In Chapter 12, Carl Hodges describes a shrimp growing venture at Puerto Penasco on the Gulf of California in Mexico's Sonora state, where more than 50 tons of whole-animal product (retailing at over $5/pound) are produced per acre/year. The enterprise, funded by Coca Cola Company and F.H. Prince Company, may just be achieving the highest value crop per unit growing area in the world.

Food from Fermentation

The simple process diagrammed in Fig. 12 is a basic system for growing food protein essentially independent of climate and weather, by means of carbohydrate or hydrocarbon fermentation. Several years ago when petroleum was much less costly, a number of major ventures were launched which grew as much as 100,000 tons a year of single-cell protein on this substrate. Today the big hope for producing a significant amount of the world's protein by this method lies in the use of carbohydrate or cellulosic plant (biomass) materials as the carbon source needed to grow the proteinaceous microorganisms. While the hydrocarbons are a richer and more efficient substrate, their fermentation requires much more oxygen

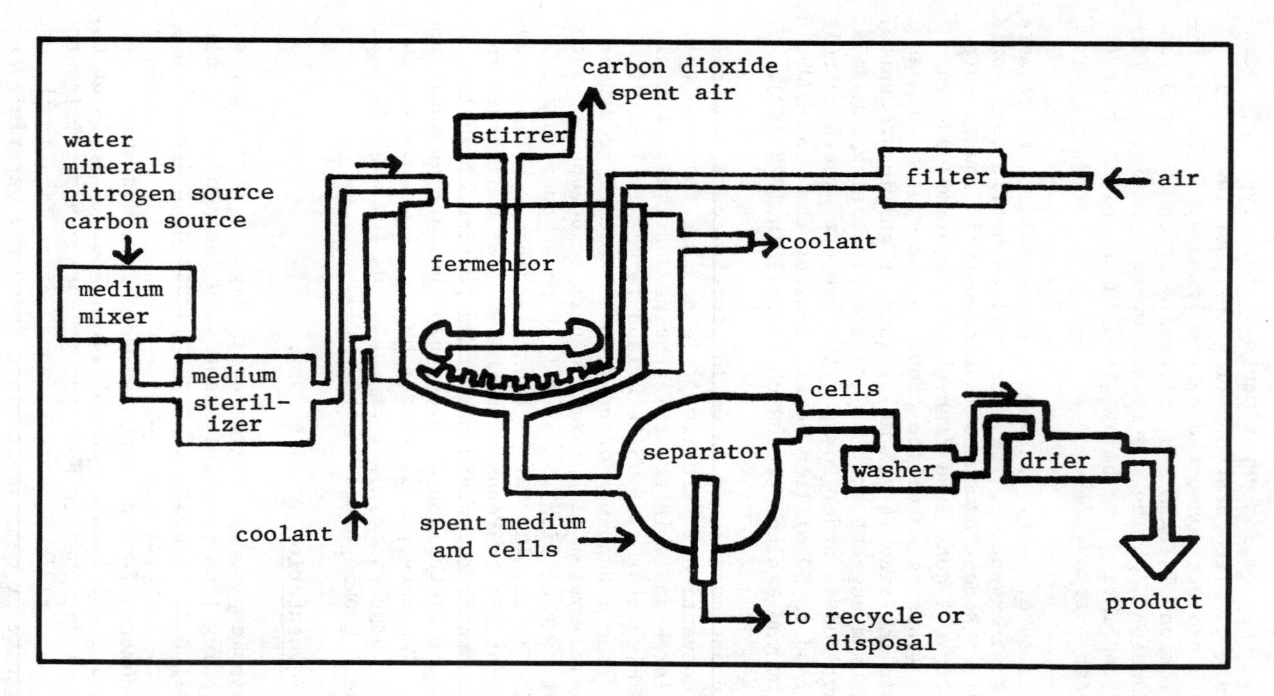

Fig. 12. Diagrammed are the elements in the process for growing food protein essentially independent of climate and weather by means of carbohydrate or hydrocarbon substrate fermentation. Source: L.E. Slater, 1976.

than carbohydrate materials; hence, significantly more energy input (23).

While the previously described aquatic farming systems suggest vast gains in productivity over conventional land agriculture, the possibilities when growing microorganisms are even more astonishing. One hundred thousand tons per year of single-cell protein plants can be grown on a few acres of land, producing as much protein as that extracted from 300,000 acres of soybeans, or from as many beef cattle as could be grown on five million acres of good grazing land (24).

The future of food production through fermentation depends very much on the economics of collecting and delivering critically-sized lots of field and crop wastes, as well as crop residues, as substrate material. The quantities of crop wastes available worldwide are impressive; there are about 58 million tons of wheat chaff, 30 million tons of corn cob material, 83 million tons of sugar bagasse, 9 million tons of molasses and vast quantities of waste from big food-processing industries, such as starchwater and peelings from potatoes, organic solids in corn and soy waste streams, and trim and peelings from fruit and vegetable canneries.

Foods Through Chemical Synthesis

The most advanced (and controversial) notion in climate-defensive food production technology is to manufacture food products on a massive scale in factories. The concept has already been well demonstrated. For instance, the chemical reaction shown in Fig. 13 describes how Germany manufactured 100 million kilograms of edible fat during World War II, starting out with coal. It is a vivid example of how, when conventionally grown food is scarce, a synthetic food or nutrient can be made in a factory by a chemical or biochemical process from nonliving raw material.

All of the basic nutrients of food -- proteins, fats, carbohydrates, vitamins -- can be produced by synthesis, and balanced, attractive, edible products fabricated by modern food engineering. Synthetic vitamin production is already a worldwide industry. Some of the important micronutrients in protein, the amino acids, are also being produced commercially by both synthesis and fermentation. Methods to build whole proteins from these constituents have been developed, but are not as yet practical for manufacturing purposes. Techniques for the synthesis of edible sugar, the basic food carbohydrate, have also been developed, and its economic manufacture (if desired) is predicted by 1985. Meanwhile, a number of high-energy substitutes for sugar have been

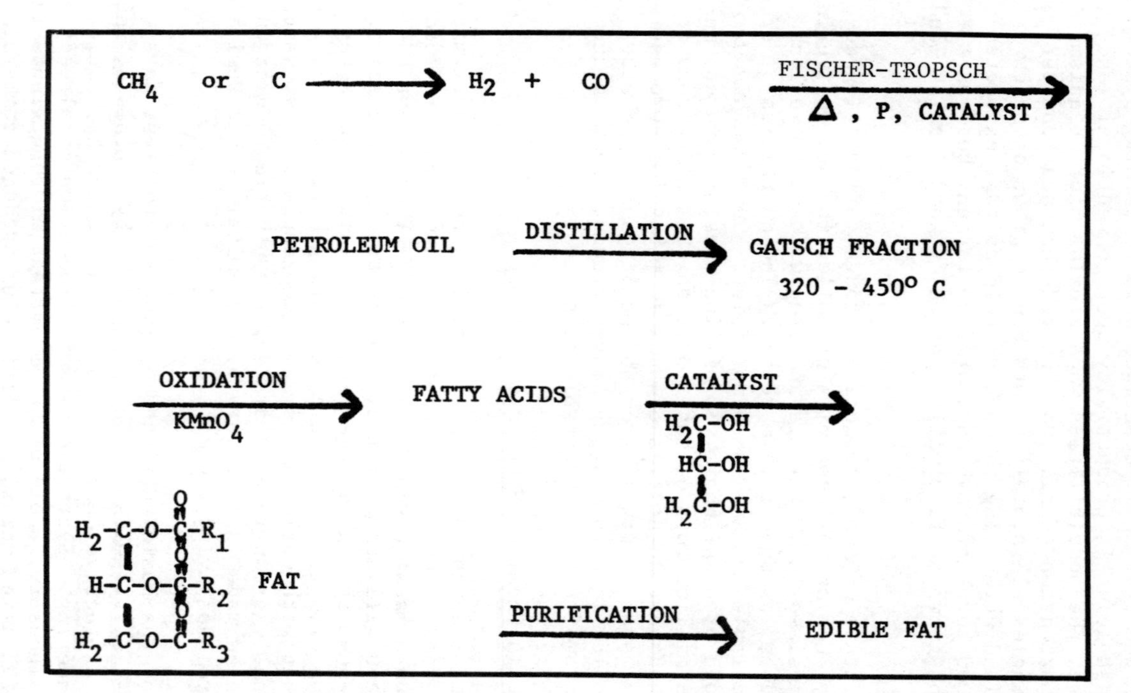

Fig. 13. This chemical reaction describes how Germany manufactured 100-million kilograms of edible fat during World War II starting from coal.

synthesized and only await FDA-type approval before widespread use.

With significant synthetic food production already commercially underway, many believe the present production of low-volume, high-unit-value items will expand naturally to have some impact on the world food supply by the year 2000. However, if a conscious world effort in R&D were to be mounted to achieve economic production technology and necessary social acceptance, large-scale climate-invulnerable food synthesis could furnish an important and stable part of the world food budget in the century ahead. An extensive discussion of the present status and potential of synthetic food production is presented by the author in Chapter 13.

References

1. Reihl, H. Introduction to the Atmosphere. 3rd Ed. Mcgraw Hill, 1978.
2. Kovda, Victor. The world's soils and human activities. In N. Polunin, ed., The Environmental Future. MacMillan, New York, 1972.
3. United Nation Environment Programme. State of the Environment Report 1977.
4. Richardson, E.V. and W. Clyma. Egypt's High Aswan Dam -- progress or retrogradation. Unpublished paper, Egypt Water Use and Management Project, Colorado State University, Fort Collins, 1979.
5. National Academy of Sciences. More Water for Arid Lands. Washington, D.C., 1974.
6. C.D. Gustafson, et al. Drip-irrigation worldwide. In Proceedings and International Drip Irrigation Congress, University of California, San Diego, 1974.
7. Kock, W. Ecological farming in the tropics. In S.K. Levin, ed., Food and Climate Review 1979, Food and Climate Forum, Aspen Institute, Boulder, 1980.
8. Kohl, D. Energy intensiveness and economic performance of livestock/crop production -- a comparison of organic vs. conventional farming systems. Food and Climate Forum, Aspen Institute/Winrock International Pilot Workshop, Morrilton, Arkansas, 1979.
9. Jennings, P.R. The amplification of agricultural production. A Scientific American Book, W.H. Freeman and Co., San Francisco, 1976.
10. Rosenberg, N.J. and K.W. Brown. Measured and modelled effects of microclimate modification on evapotranspiration by irrigated crops in a region of strong sensible heat advection. Proc. Symp. on Plant Response to Climatic Factors. UNESCO, Paris, 1973. Ecology and Conservation, Volume 5, 1973.

11. Wittwer, S.H. Environmental and societal consequences of a possible CO_2-induced climate change on agriculture. Annual Meeting of the American Association for the Advancement of Science, Washington, D.C., 1980.
12. Somers, G.F. National halophytes as a potential resource for new salt-tolerant crops: some progress and prospects. In A. Hollaender, ed., The Biosaline Concept. Plenum Publishing Co., 1979.
13. USDA Quick Bibliography Series. Weather Models and Simulation Studies Applied to Agriculture. NAL-BIBL-79-29. USDA National Agricultural Library, Beltsville, Md., 1979.
14. Lofchie, Michael F. Political and economic origins of African hunger. Journal of Modern African Studies 13:4, 1975.
15. Ibid.
16. Basse, M.T. We are trying to make local products as attractive as imported ones. Ceres, May/June 1978.
17. Spurgeon, D. Hidden Harvest: A Systems Approach to Post-harvest Technology. International Development Research Centre, Ottawa, 1976.
18. Slater, L.E. A Global View of Food Processing Engineering. Proceedings of the Second International Congress on Engineering and Food, Elsevier Scientific Publishing Co., Amsterdam, 1980.
19. The Role of Ruminants in Support of Man. Winrock International Livestock Center, Morrilton, Arkansas, 1978.
20. General Mills big gamble in indoor farming. Duns Review, November 1979.
21. Fish Can Grow Year-Round in Power Plant Effluent. Texas Agricultural Experiment Station Research Report, College Station, Texas, 1976.
22. Aquaculture in the United States: Constraints and Opportunities. National Research Council/National Academy of Sciences, Washington, D.C., 1978.
23. Slater, L.E. Food waste fermentation: a special report. Food Engineering International, Radnor, Pennsylvania, October 1976.
24. Malick, E.A., et al. Single-cell protein: its status and future implications in world food supply. Second Arab Conference on Petrochemicals, Abu Dhabi, 1976.

James D. McQuigg

8. The Promise of Food and Climate Information Systems

Introduction

The concept of "contending with climate" conjures up an image of man defiantly facing nature as an adversary. According to Webster's New Collegiate Dictionary, "contend" is defined as "to strive or view in contest or rivalry against difficulties." Given the continuing, ever-present impact of climatic change and what farmers must do to be successful, the image seems a likely one.

Year-to-year variability of climate requires planning agricultural activities a season or more in advance. It necessitates constant adjustments in the financing, organization and management strategy for a farm, agricultural business or government agency. And even if climate were serene and steady, man would still have many challenging problems in the design and management of food production and distribution systems. The uncertainty of international events, political changes, and rapidly advancing agricultural technology have put pressures on agricultural management, adding strongly to those induced by climatic variability.

Interest in climate and climate impact has intensified due to the series of major crop failures in the 1970s. Continuing increase in world population, together with rising expectations for a better standard of living, has put additional pressure on food supplies, and at times has increased the magnitude of the impact of climatic variability on food production and distribution systems. At the same time, many significant improvements in information-processing technology, with their promise as techniques for contending with climate, have coincided with this same series of years.

There has been considerable effort over the last three or four decades to reduce food shortages and nutritional stress. Prominent measures have included large food shipments, introduction of high-yielding varieties of wheat and rice, expansion of irrigated areas and of electric power distribution facilities, improved roads, food storage and transport systems, and the growing presence of agricultural scientists and technicians in developing countries. Similar efforts have increased food production and improved the distribution of food in most of the developed (food exporting) countries of the world.

The idea that food and climate information systems should be included in a list of agricultural development tools is not quite as controversial today as it was five or ten years ago. But the development and actual application of such systems is just starting to become a priority item. Rather than extend the argument that food and climate information system development should receive more attention and support, this chapter will promote their cause by describing some successful applications, together with a few examples of how applications should not proceed.

Generating the Information

There are few physical limitations on the production and transmission of weather and climate information. The weather services of the world collect and communicate a very large volume of observations of conditions at the surface of the land masses and oceans and in the upper atmosphere. In addition, there are observations from surface-based radar and from satellites in space. Most of this information is relayed very quickly from one meteorological center to another.

It is true that there are times when the meteorological communicating systems experience difficulties because of physical failures or the effects of magnetic storms. It is also true that there are times when it would be desirable to have observations from a more dense network of reporting stations in certain regions of the world. Despite these shortcomings, it is very rare when a major weather system develops and moves through any part of the world without being detected by at least one meteorological center.

There are also good historical climate data available for most agriculturally productive regions of the world. Such historical data series often cover more than 30 years. Again, it is true there are parts of the world where it would be desirable to have either longer records or more numerous sample locations. Despite this, however, it is possible to produce reasonably good estimates of the variability in space and time of the climatic events impacting on production and distribution of food for most agricultural regions of the world.

Producing weather and climate information by itself has little or no direct effect on the outcome of a crop or on the outcome of distribution once it is produced. The value of weather and climate information comes through its use in decision-making by those who manage agricultural systems. Fishburn (1964) delineates the problem of information use in the following remark:

> "Since the origin of the species, men have been making decisions, and other men have been telling them how they either make or should make decisions. The scientist may be better prepared and able to deal with the analysis and synthesis of complex organisms and systems than is the manager who must act on the information generated by scientists and technicians. Managers have been careful to insure and protect their decision-making prerogatives and, for the most part, management scientists have been just as careful not to infringe upon these prerogatives."

Some Applications

Aspen Institute's Food and Climate Project

In June 1978, the Food and Climate Forum, a program of the Aspen Institute for Humanistic Studies, decided there should be an attempt to demonstrate that a food and climate information system could be employed in improving food production and nutrition in a Third World country. Rather than simply support a position paper on the subject, Aspen Institute opted to test its subject through an experiment in the field, and thereby perhaps develop a model for implementation and similar effort elsewhere.

After consulting with specialists and agencies who had long experience with food systems, both in the developing countries and in more advanced agriculture, Venezuela was selected as a likely site for the project. Further discussions with officials in the Ministry of Agriculture in Venezuela led to the conclusion that the proposed information system might best be tested by devoting it to the problem of potential expansion of corn production in a newly-opening farm region in the Western Llanos, or savannah region of the country.

The llanos region of western Venezuela is potentially

attractive for significantly adding to the nation's domestic production of corn, both for human consumption and feedgrain. It has traditionally been used as rangeland for cattle, with some crop production now underway. Its eastern and southern portions are very wet and swampy, forming the headwaters of the Orinoco and Amazon basins. The western fringe of the region is just east of the Andes, with temperatures too cool for corn. Somewhere in between the high country and the Amazon-Orinoco headwaters there are sections in the Western Llanos that have a soil-climate combination that is most suitable for expansion of Venezuelan corn production.

Selecting and evaluating the best locations for growing corn in a region with wide climatic variability and myriad soil types is not a simple task. A survey of just the soils in the region would be helpful, and in fact, such a survey has just recently been completed. Analysis of the climatic averages alone of particular locations would also be helpful. But soil or climate information by itself would tell only part of the story.

Further consultation with Venezuelan climatologists and agronomists has led to the development of a soil moisture model for the selected study area in the Western Llanos region. This model, under development at the time of this writing (March 1980), will be applied to a reasonable historic sample of climatic data. The result will be a synthesis of the best available information concerning the soils and the year-to-year variability of climate of this region. Interpretation of the information will be based on crop modelling techniques that have been used successfully in a number of major crop-producing areas of the world.

The final product in this experimental model-building effort will be in the form of a simulation of moisture stress on a crop that has not yet been grown on a large commercial scale in the Western Llanos. There is as yet no good series of information on corn yield and production in the area. But the simulation model results will furnish information that can be used to identify three major categories of sub-regions in the Western Llanos:

1. An area where the soil-climate combination is most suitable for corn production.
2. An area where the soil and/or climate are definitely not suitable for corn production.
3. An area where corn production will succeed only if adequate irrigation is finished.

If the application of the soil moisture stress simulation model is successful, it will be because policy level officials of the Venezuelan government and the farmers of the Western Llanos accept the information produced and carry out an effective expansion of corn production in the area. It is likely the effort will succeed, however, because the experimental work was preceeded by careful discussions with officials and scientists in Venezuela who understood the problem at hand well enough to perceive that the proposed technology might be able to contribute a solution. In effect, they also essentially contributed to the design of the experiment itself.

Misdirection in the Sahel

The African Sahel has been a chronic food problem region for hundreds of years, supporting only marginal agriculture. In the 1950s and 1960s there were many agricultural development projects, funded by the United States and European governments, as well as by one or more United Nation agencies. These projects were designed to increase the production of grain or improve the management of livestock in the Sahel. A recent article in The Economist (June 11, 1977, pp. 78-79) includes the following:

> "...ironically, it was the misconstrued development schemes, in which boreholes were drilled and cattle vaccinated, that most contributed to the environmental disaster. Around the wells, which were too few and scattered, a great press of animals ruined the pastures and eroded the sub-soil....a succession of good years disguised the incipient calamity. Only a temporary deterioration in the climate was required to bring the whole system to its knees. The effects of well-intentioned experts from the U.N. Food and Agriculture Organization had disrupted the precarious balance of the traditional land-use system."

An adequate sample of historical climate records is available for enough stations in the Sahel to clearly show that recurrent periods of drought are quite likely, and to show that the period of years during which improved agricultural technology was brought to the region (1950-1969) was an anomalous run of favorable rainfall years.

What happened in the Sahel is a classic example of an attempt to make a region something it could not become because of climate constraints. It was a case of information on food and climate being available, but unheeded.

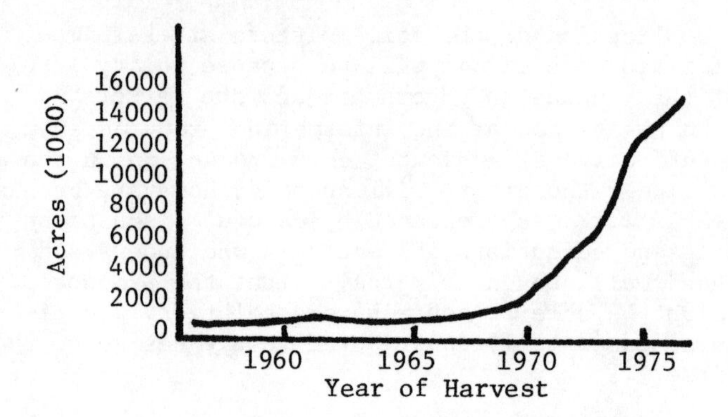

Fig. 1. Brazil Soybeans: Area Harvested.

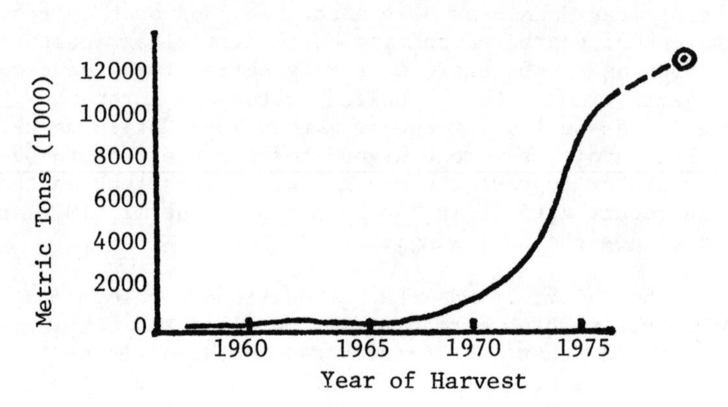

Fig. 2. Brazil Soybeans: Production.

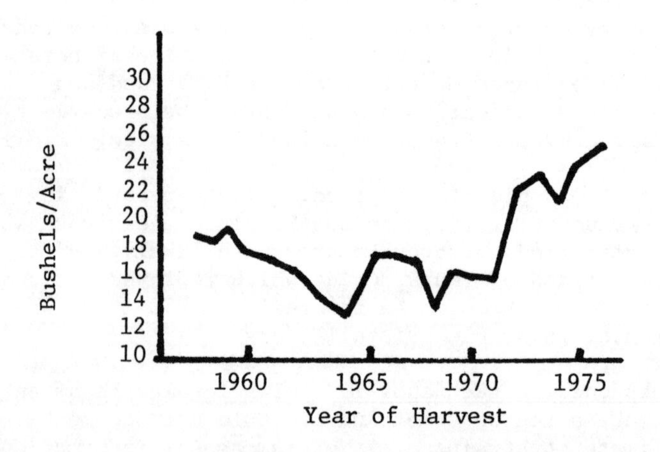

Fig. 3. Brazil Soybeans: Yield.

Soybeans in Brazil

Recent increases in the production of soybeans in Brazil offer another example of the application of modern agricultural technology to meet the demand for food in a climatically difficult region.

Following the rapid rise of world grain prices in 1972, the Brazilian government encouraged a very large expansion of soybean acreage. New seed varieties were introduced, together with the use of herbicides, insecticides and larger machinery. The results of this effort are illustrated in Figures 1, 2 and 3.

The Brazil soybean harvest of 1978 was projected to be near 12 million metric tons. Low rainfall in October and November of 1977 resulted in a serious delay in planting. Very heavy rains came to parts of the Brazil soybean region in December 1978, resulting in the need for some replanting. Then followed a very dry January and February, right at the time the crop reached its peak need for soil moisture. There were widely varying estimates of soybean production in the press, ranging from seven to 11 million metric tons. These inaccurate estimates resulted in a highly volatile soybean market, with large fluctuations in the price of soybeans and other grains.

While the historical series of yield and production values for Brazil's soybeans is short compared to the length of record for the United States soybean crop, there is a reasonably good sample of such data, together with a series of data on temperature and rainfall. Since the 1977-'78 crop season we have developed a weather/soybean yield and production model which has been used during the ensuing two crop seasons to prepare remarkably good estimates of the size of the Brazil crop.

Weather data are available each day from a network of stations in the crop regions of Brazil (as they are for most other agricultural regions of the world). These data are not restricted, and they can be obtained by a wide variety of government agencies, grain-trading companies, farmers, etc., if they will invest the modest resources necessary.

Information about the size of a commercially important crop that is produced through analysis of a flow of weather information does not add one gram of additional grain to the supply available in the world. But it does add some degree of stability to the price situation. Countries that are short of food and have to manage an active import program can use weather-based crop estimates to time their purchases. Companies that buy grain and convert it into food products can stabilize their costs and remain competitive in the prices they charge.

Monitoring Internal Food Production

A food and climate information system, if properly used, can yield important savings to a chronic food problem country willing to invest scarce resources and allocate its limited scientific-technical manpower to its development and operation. The system would be designed to monitor the flow of crop-season weather within the country, and within the countries from which they purchase the grain needed to supplement their own production. There is no doubt that this would furnish timely, reasonably accurate estimates of the impact of weather on the availability and price of grain. Of course, the results of such an effort must be well-communicated to officials in the government responsible for internal distribution of food and management of grain imports. The result would be better timing of grain imports, when prices are most advantageous, and reduction of food supply distribution gaps due to unanticipated local shortfalls.

The problem is not how to develop such an information system. Rather, it lies in convincing officials responsible for food policy and food programs that funds allocated for this technology and its use will furnish a quick return and continuing future benefits.

One of the best examples of a food and climate information system at work is the one functioning within the Canadian Wheat Board in Winnipeg. Now, to be sure, Canada hardly typifies a chronic food deficient area. But its use of its system furnishes a model worth study by all.

The Canadian Wheat Board has the responsibility of marketing most of the grain produced in Canada. It has a long record of success in world-wide marketing, much of it due to an ability to monitor the impact of weather on Canadian production and world production -- a capability which developed rapidly following the now famous 1972 failure of the USSR grain crop and other weather-induced production shortfalls that same year.

That capability developed in the following manner. The Canadian Wheat Board first asked Dr. George Robertson, an eminent agro-meteorologist, to consult with them on the problem. He recommended that the Board should increase its ability to monitor the flow of daily weather data from most of the important grain-producing regions of the world. At first this resulted in their having more "raw" data than they knew what to do with. But in another year or so the in-house computer capability of the Wheat Board was increased, and some of the staff were given the job of setting up a program to monitor the impact of weather on a global scale.

Then, about two years ago, the Canadian Wheat Board hired Dr. John Benci, who received his graduate training in meteorology and agronomy at the University of Missouri. He has since provided strong technical leadership in the information system program.

Acceptance of the idea of monitoring weather impacts on grain production did not come to the Wheat Board overnight. The important factor is that officials of the Board did perceive the need for such a capability, and that they proceeded to establish their own program to meet that need. To the best of present knowledge, the weather-monitoring section has been fully integrated into the world-wide marketing staff of the Wheat Board, with daily contacts adding continuously to their credibility.

It would be foolish, of course, to recommend that every chronic food problem country should duplicate the sophisticated monitoring capability of the Canadian Wheat Board. But the advantages the system offers to Canada in its trading on the world grain market have been well demonstrated. It can thus be argued that grain-buying countries that do not have such a capability will continue to pay more for their purchases than might be necessary.

One of the critical problems is that there just are not enough well-trained agricultural meteorologists around. If as many as ten countries were to decide to hire one or two such persons, they could well dry up the supply.

Some Further Examples

Irrigation systems have been in use for centuries in many of the arid and semi-arid regions of the world. One is now beginning to see a significant increase in use of supplemental irrigation systems in semi-humid climatic regions; e.g. in the U.S. Mississippi Delta, and in the Corn Belt. In a region like the Corn Belt the average annual rainfall is quite sufficient to support efficient corn production. But there are some years when nature does not deliver enough rain at the right time. The problem is no longer one of convincing farmers they should use supplemental irrigation. The problems are:

A) Choice of the particular system;

B) Financing the installation;

C) Managing the schedule of irrigation applications to minimize cost and maximize the impact on yield.

There is an information system called "AGNET" which was developed at the University of Nebraska to deal with the scheduling of supplemental irrigation. The system provides access to a central computer data bank through interactive terminals located at the Extension Service Offices in about seven states. A farmer can enter data on his own fields and and on the rainfall and temperature in his locality, and receive back information on the soil moisture situation and recommendations for the timing and amount of irrigation he should apply.

The same "AGNET" system can be used to obtain information concerning the management of a beef herd, the drying of grain in storage, and other farm operations. A similar computer network (called "FACTS") has been developed at Purdue University to serve Indiana farmers.

A study of Georgia weather records by Skinner (1978) shows that in five out of ten years there are 60 to 70 drought days in the central part of the state, and 50 to 60 days in the southern third. When enough of these drought stress days occur at the peak of the growing season for crops like corn, soybeans and peanuts, there is a very large drop in yield. Clearly, a well-managed program of supplemental irrigation could offer important savings in this region.

The problem of scheduling irrigation in a humid or semi-humid climate region, however, is much more complex than scheduling in arid-region farming, because of the vastly higher probability that naturally occurring precipitation will arrive. When moderate to heavy rains fall right after an irrigation application, there is a risk that yields will actually be depressed. There is also the obvious loss in wasted energy and labor.

A quantitative approach to weather impact and related factors is becoming a way of life for more and more farmers in the United States. Many have begun to use programmable calculators, and others have bought mini-computers. They are either learning to write their own programs, or are buying programs and software from consulting firms or from university extension service offices.

Conclusions

The central theme of this discussion has been that weather and climate variability, despite great progress in agricultural technology, still has a powerful influence on yield, production, cost and demand for food. Information on the climatic factor, an essential factor in agricultural management and planning, is now available on a world basis in surprising breadth and low cost.

When we talk about information which is concerned with food production and agriculture we need to be thinking in terms of events that happen on particular farms; on the management of land, machinery, labor and other factors that are under the control of farmers. We also need to consider a broader system of information that includes the management of organizations that provide materials and services to farmers, plus the governmental and academic organizations that provide services (or perform regulatory functions). In addition, the problems related to food systems are increasingly global in nature. A drought in Brazil, India, China or the USSR has a worldwide impact on grain price and demand.

The promise of an effective food and climate information system is to improve the management of production, processing and distribution of food. To the extent that climate has an impact on one or more links in the food system, there is a potential need for information about this impact, especially if it can be delivered in a timely, understandable format to the decision-makers.

References

Fishburn, C. Decision and Value Theory. John Wiley and Sons, New York, 1964.

Skinner, R.E. Georgia's fastest growing area. Irrigation Age, July-August 1978.

Howard E. Worne

9. A Microecological View of Arid Land Agriculture

Introduction

Much of the land on our planet, approximately 30 to 35 percent, is located in hot, dry regions. An area of this magnitude should to all intents and purposes have vast potential for profitable agricultural production. But under present conditions, is this possible? What would be the economic and environmental costs in developing such an area most effectively?

It has long been held that water is the limiting factor in arid lands development. Such a belief, however, is no longer widely accepted by knowledgeable agronomists. Many arid areas, even with water, are unable to obtain desired productivity because organic matter in the soil and its microflora have been reduced by climate impact and unfavorable human activity to suboptimal levels. Besides their lack of a crop-sustaining microecology, such soils retain water minimally, and irrigation is required frequently and at correspondingly high cost.

Restoring dry, impoverished sandy soils requires the addition of organic matter and rebuilding of soil microflora through the use of special microorganism innoculums. Given the benefits possible, the costs in this approach to arid land rehabilitation are miniscule, especially when compared to the huge expenditures required for new irrigation schemes. This chapter reviews the problems of arid land soil management, the contributing role of microflora, and the promise of new technology in selected and improved soil microbial additives.

The Arid Land Problem

Constant exposure of soils to strong winds and intense

solar radiation, with all of their deleterious effects, will always be a problem no matter how much water is available. Moreover, a high percentage of arid land, besides being extremely low in organic matter and essential nutrient reserves, is either too shallow, rocky or steep for productive farming and efficient mechanical management. Salinity also prevents much land from being used for anything except producing salt-tolerant flora and fauna.

The irrigation of alkaline and calcerous soils containing low levels of organic matter in no way changes them into highly productive soils. At best, abundance of water might convert some selected areas of the arid regions into oases. But distinct changes in the normal physical environment, as a result of the increased water, are usually accompanied by significant alterations in the local ecosystem. This in turn creates an entirely new set of problems.

Abundant use of irrigation water, unless properly controlled, creates wastewater disposal problems, and it is important that adequate drainage systems are installed to manage the outflow. The irrigation canals which supply water to agricultural areas also furnish passage for various disease-causing fauna. Both insects and infectuous microorganisms thrive in the more humid environment, causing potential health problems in heavily populated areas. It is thus essential, prior to making any major environmental changes, that the overall parameters of the proposed changes are carefully studied so that solving one problem does not foster several others.

Arid lands in the Middle East are an interesting case in point. At present they are in a state of marked transition. Once considered of limited importance, oil extraction and increasing technological capability have established potential for improved agriculture and expanded populations, which in turn could foster additional economic strength. The philosophy of wanton exploitation of resources is still practiced, but an awareness of the special problems of arid land development is rapidly spreading. It is now realized that uncontrolled exploitation comes only at a price, and that price can be exorbitant.

Origins of the Problem

Most desert regions in the Middle East and throughout the rest of the world have developed as a consequence of long-term climatic changes, often aggravated and enlarged by over-ambitious farming, grazing and deforestation. Comparing the ancient deserts of Asia and Africa with arid and semi-arid areas of the Western Hemisphere reveals a similarity in origins and problems. In essence, the arid and semi-arid regions of the Western Hemisphere provide an excellent scenario on the destruction of potentially productive land by indiscrete farming and improper management.

Studying the Sahara Desert, a vast region which occupies a land area almost equal in size to the United States, reveals that both extensive climatic changes and secondary human influences were involved in its formation. The Sahara's present aridity probably developed as the result of significant alterations in the flow of high-altitude tropical windstorms, indicating long-term, rather than short-term, geological activity (1). It is also suggested that during the late Pleistocene era there were four main pluvial periods when the Sahara was a productive agricultural area with adequate surface waters (2). There is ample evidence, moreover, that during the Mesolithic Age, the climate of North Africa was considerably more humid than at present, and that this continued until the onset of the Neolithic period. Local populations raised cattle and other domestic animals, even in the northern Sudan, which has now achieved desert status. This period of adequate moisture, which appears to have terminated about 3,500 years ago, is believed to have coexisted with a similar period in East Africa (3).

It is also thought that random climate changes led to eliminating the original Sahara fauna and flora, unable to adapt after the last pluvial period. As a result, deterioration of the Sahara region has taken place, which in turn has been greatly accelerated by irresponsible human involvement (4). Whichever assessment is correct, there is no doubt that climatic conditions have changed considerably in these areas since the Pleistocene period. Until relatively recent times the biota of the central Nile Valley was still rather abundant. The decimation of wild animal life by destructive and wanton excesses of the Roman Empire did much to worsen the effect of climatic change, and aided in the annihilation of significant segments of the wildlife (5). Unfortunately, it is not possible at this late date to accumulate acceptable data which substantiates the pattern of climatic changes that have been developing over many thousands of years. But it is clearly demonstrated that the rate of destruction of the flora and fauna has greatly accelerated in recent years.

Data suggest a significant portion of the Sahara Desert developed between 10,000 to 20,000 years ago (6). In direct contrast, much of the desert area of Arizona and New Mexico evolved from short-term climatic changes coupled with excessive farming and grazing over the last two centuries. Flora and fauna now found in this region migrated over a period of time from the desert regions of central Mexico, formed over a million years ago as a result of long-term climatic changes. The suggested migration helps explain the high degree of specialization observed in plants and animals found in Arizona and New Mexico. One can assume a variety of problems culminating in the creation of arid desert lands. Although major changes in climatic conditions were foremost in most cases, man unfortunately has contributed to enlarging these desert areas.

Some Developmental Factors

Developmental projects in arid land areas must be established under circumstances totally different from those practiced over many generations. It is safe to assume that no native population, under present world conditions, will be satisfied to remain in arid areas in the future under circumstances which promulgate their present low economic standards. The general immigration trend, especially among the youth, from the backward areas to more populated and progressive places, further retards development in the areas from which immigration occurs. Unless viable economics based on oil income or modern agriculture can be developed, the more progressive among the population tend to move where opportunity exists.

Under present economics, arid zone agriculture must utilize some form of irrigation to produce high value crops in areas distant from available markets. Since production costs are higher than normal, it is important for farmers to choose crops with high unit value. Vegetables and fruits grown on irrigated land, rather than maize, wheat and barley, are the most valuable arid land crops today.

At the same time, regional irrigation, like any modern development, requires an intricate supporting infrastructure. Involved are surface transportation and electric power, as well as other services such as schools, hospitals and investment opportunities to support an expanding population. Until now, the dispersed settlement pattern of arid land has been a marked hindrance to establishing a more functional and economic regional infrastructure. The cost of such an infrastructure under present economic conditions makes consolidation of agricultural areas mandatory. This consolidation of agricultural activity in selected regions maintains the typical arid land pattern of partition, with concentrated occupancy and cultivation separated by extensive unoccupied zones (7).

One of the major exceptions to land concentration involves crops such as vegetables and fruits which efficiently utilize air transportation for shipment to major consuming areas. In essence, the distance between the agricultural area and the airport does not materially affect the cost of infrastructure development. Compared to more humid areas of the world, arid regions also have a higher percentage of fog-free weather, resulting in more flying hours. One major exception to this climatic bonanza exists in the west African coastal deserts which, because of environmental conditions, experience severe operational limitations during those seasons when fog is a problem.

Irrigation Problems and Techniques

Excessive aridity, coupled with a disproportionate mineralization of the soil, causes many arid zone soils to become saline. To reduce unwanted mineralization in irrigated arid zones, adequate drainage is essential. Hence, substantial additional investment is usually needed to adequately develop agriculture in the arid zones. Moreover, a high degree of expertise is required in initial planning and implementation to curtail the excessive water consumption which leaches minerals and creates salinity problems.

The arid land farmer must become more knowledgeable in water management than his humid zone counterpart, for he must know more precisely when to apply the proper quantity of water for optimum results. He faces an excessive rate of water evaporation from winds and high temperatures, which can significantly reduce overall benefits. Larger quantities of water are needed for arid areas than for humid regions to obtain comparable results.

The prospects for more efficient water application in arid land farming have been significantly advanced by a highly practical system called trickle irrigation, which employs pipes laid on or under the ground. Valves, fitted to the pipes at spacings chosen by the farmer, release predetermined quantities of water, which can also contain required plant nutrients, directly to the plants for calculated periods of time. The mechanical layout thus delivers water and nutrients in quantities which can be absorbed at the time of application, thereby vastly improving effective utilization over conventional practice. Farm experiments with selected vegetables have given yields that in some cases doubled those obtained by standard methods of irrigation (8). The use of controlled trickle irrigation permits the use of liquid fertilizers, including novel concentrates such as hydrolyzed water-soluble fish proteins. Thus, the usual problem of fertilizer loss due to uncontrolled biological decomposition or leaching is eliminated and the cost per unit reduced considerably.

Underground aquifers have been the primary source of water in most desert regions. Reserves of ground water in these regions are possibly the remains of water retained from earlier pluvial periods, either in the unconsolidated mantle or in the consolidated water-bearing rocks (9). It has been suggested that the latter are recharged periodically by occasional rains.

Achieving significant utilization of available ground water in most cases requires widespread underground drilling for fossil waters. In some areas such supplies may be quite extensive. A comprehensive analysis of the ground water in Libya well illustrates the various types of supplies that may exist. In this dry country, only on the coastal plains near Tripoli, northeast of Jabal Nafusah and Bengazi and south of Jabal Al Akahbar, is there sufficient rainfall to recharge the ground water each year (10). Population studies show that approximately 90 percent of the inhabitants live in this more humid area. It has been estimated that approximately 10 percent of the country receives rainfall in excess of two centimeters per year. Geological exploration in the interior has demonstrated that good quality water can be obtained from the widespread Nubian sandstone and related aquifers. In addition, a diminishing groundwater supply south of Jabal Nafusah provides needed oases irrigation with residual water that possibly accumulated during the Ice Age.

In many instances, the desert ground waters in Saudi Arabia, Kuwait, the United Arab Emirates and other arid countries are likely to be of poor quality because of high salt concentrations. A significant portion of the available ground water contains as much as 1,200 parts per million (ppm) of sulphates, 1,500 ppm of chlorides, or a total of 3,500 ppm of total dissolved solids (11). Chemical analysis has shown this highly mineralized ground water contains more than four times the salt content of seawater. In the Kufra region of Libya, which lies in the southeastern part of the country, there exists a 140,000 square kilometer reserve of good quality water which estimates suggest should be able to irrigate approximately 1,200,000 acres of productive agricultural land for the next 800 years (12).

The Cooperating Microbiota

While water and soil are essential components of successful arid land agriculture, too little attention has been paid to the soil microorganisms, a group of microbiota crucial in promoting plant growth. Analysis of arid soils from various parts of the world reveals a common finding: the level of soil microorganisms is extremely low. In humid area agriculture, soils contain as much as several billion live microorganisms per gram, while in dry arid soils less than 50,000 microorganisms per gram are found, and these are essentially spore-forming bacteria, Actinomycetes, fungi and algae (13). Table 1 illustrates the comparative distribution of the microorganisms from various regions.

To fully appreciate the importance of microorganisms for productive agriculture, it is first necessary to understand the types involved and the function of each. Soil contains myriads of microorganisms, which are truly the housekeepers of the earth. As the result of a totally interrelated balanced microecology, they are able to convert, through a great variety of processes, the plant, animal, bird and insect debris of one season into the nutrients of the next. This complex biological enterprise is accomplished by the interaction of symbiotic groups of microorganisms. Species whose activities are interdependent and mutually beneficial secrete a broad complex of enzymes necessary to catalyze synthesis and waste decomposition.

Various types of specialized microorganisms, operating somewhat miniature chemical plants, make up the soil microflora. The photosynthetic algae found in the soil, including the green Chlorophyceae, blue-green Cyanophyceae and Diatomacea, are unique because they possess chlorophyll, and are thus able to synthesize carbohydrates from sunlight and carbon dioxide, giving off oxygen in the process. This oxygen is most essential to other members of the microecosystem, for many of the most active microorganisms are aerobic and unable to function with any degree of efficiency in the absence of oxygen. Algae themselves, because of their metabolic structure, have little or no effect in decomposing natural polymeric molecules in the soil, such as cellulose and lignin. Rather, they delegate this function to the non-chlorophyll bearing fungi, which are highly organized life forms capable of diverse biochemical activity.

The soil fungi include a large group of organisms known as Phycomycetes, Ascomycetes, Hypomycetes and Basidiomycetes. Due to their incredible biochemical activity, these fungi are able to secrete a vast array of enzymes capable of

Table 1. Distribution of Microorganisms in Various Soils (microorganisms/gram of soil x 10^3).

Location	Soil pH	Aerobic Bacteria	Anaerobic Bacteria	Actinomycetes	Fungi	Algae
California	7.2	8215	1650	2245	155	30
New Jersey	7.1	7980	1790	2110	122	26
Louisiana	6.5	8100	1824	2000	145	38
W. Texas	8.2	1015	930	775	80	11
Arizona Desert	7.8	15	10	9	11	1
Mohave Desert	7.2	18	9	11	10	2
Abu Dhabi, U.A.E.	8.6	20	7	10	8	0
Saudi Arabia	8.2	17	6	9	4	0
Kuwait	8.0	19	8	11	5	1

biodegrading a wide variety of substrates ranging from low molecular weights to complex materials such as cellulose and lignin. Under proper conditions fungal vegetative growth throughout the soil can be so extensive that massive particles are held together by means of a very fine microscopic network of mycelium and metabolic by-products.

On the other side of the soil microecosystem we find the bacteria, a rather large group of small microorganisms with a relatively primitive cellular structure. They include both spore-forming and non-spore-forming rods, cocci and spirilla. Bacteria vary considerably in size, oxygen requirement, energy utilization, bio-gel formation and biochemical activity. Although this group of microorganisms has the ability to secrete various types of enzymes, many are limited and are not able to secrete cellulose and pectin-digesting enzymes necessary to degrade the more complex natural polymers such as cellulose.

The Actinomycetes are intermediate, developing between the primitive bacteria and the more highly developed fungi, and sharing some of the cultural and biochemical attributes of each. Like the fungi, these microorganisms show a higher degree of differentiation and ability to secrete the complex enzyme systems involved in both synthesis and biodegradation. Three genera of Actinomycetes are well represented in the soil. Species of Nocardia are closely related to some of the bacteria, especially the mycobacteria and the corynebacteria. Species belonging to the genera Streptomycetes and Micromonospora are more closely related in both structural and biochemical activity to the true fungi. Actinomycetes vary greatly in their biochemical properties, in their relation to high plants and effect upon soil bacteria. They display both associative and antagonistic interrelationships.

Protozoa in the soil essentially include amoeba, flagellates and ciliates, with their vegetative versus cyst structures attracting considerable interest. Interest has also focused on their relation to bacteria, since it has been suggested that protozoa function in the soil as the natural predator of its bacteria. By utilizing bacteria as a source of food, it is believe that protozoa exert a control on their abundance, thus in turn affecting a variety of soil processes (14).

The final group of microorganisms in the soil includes phages and various viruses. Although our immediate knowledge of the importance of these forms to the soil process, particularly in arid soils, is quite limited, their ability to attack both lower and high forms of life, ranging from bacteria and Actinomycetes to plants and animals, makes them of great potential importance.

Power of the Blue-Green Algae

The blue-green algae, because of their ability to obtain metabolic energy by photosynthesis rather than from agricultural residues and their close interrelationship with other essential soil microorganisms, are an important factor in building fertile soils. Blue-green algae are uniquely able to engage in photosynthesis, nitrogen fixation and growth under aerobic conditions, and therefore have special potential for enhancing productivity in arid agriculture. A major drawback of the Azotobacter species, on which so much experimental effort has been expended, is the high level of energy in the soil in the form of plant residues required to enable them to fix atmospheric nitrogen. Conversely, blue-green algae are capable of utilizing atmospheric carbon dioxide instead of plant residues as a source of metabolic carbon for synthesis and growth, with solar radiation supplying their energy needs.

Because of their special physical and biochemical capabilities, certain of the blue-green algae are the predominate flora in arid deserts. These species are found in abundance on rocks and buildings receiving total exposure to the full intensity of tropical sunlight. Other species with different biochemical characteristics, however, are only able to grow in temperate and freshwater lakes. Some are the dominant plant life in the frigid waters of the Antarctic, while others are among the few microorganisms able to grow in hot springs at temperatures above 50 degrees C.

Blue-green algae thus form a substantial portion of the biomass in several important types of ecosystems. If natural resources are to be conserved and used to optimum advantage, the role of the blue-green algae merits attention, for these exceptional microscopic organisms are of direct practical importance to agriculture. For example, nitrogen fixation by blue-green algae has made a major contribution to the fertility of rice fields in the Far East (15). In many rice-producing countries peasant farmers generally do not apply any fertilizer, whether inorganic or organic. It appears that growth of glue-green algae, either alone or in symbiotic combination with selected native flora, often contribute to a good harvest, whereas when absent only meager yields are obtained. Instead, it is reasonable to assume that millions of peasant farmers have been able to survive only because of natural nitrogen fixation by the blue-green algae.

In underdeveloped countries located both in the humid and arid semi-tropical and tropical regions of the world, a commercially-prepared microbiological culture such as Azobac, which has the ability to efficiently fix atmospheric nitrogen *in situ*, has many advantages over nitrogenous fertilizers which, after manufacture from expensive fossil fuels, must be bulkily transported to the farming areas. Research and development programs for increasing the effectiveness of algal nitrogen fixation are therefore of considerable economic importance. In addition to their nitrogen-fixing ability, blue-green algae are of significant value in helping to reduce soil erosion, increasing the organic content of soils, and producing bio-stimulant growth factors for higher plants. In order for blue-green algae and other nitrogen-fixing microorganisms to accomplish these growth functions, trace quantities of manganese, boron, molybdenum, copper, zinc and cobalt are quite essential for enzymatic synthesis.

With their rather unique biochemical capabilities, blue-green algae are able to liberate substantial quantities of extracellular nitrogenous compounds into the surrounding medium, irrespective of whether they are growing on either elemental or combined nitrogen. The amount of extracellular fixed nitrogen made available varies considerably according to environmental conditions and the strains of algae involved. Usually between 20 to 40 percent of the total nitrogen metabolized is found as extracellular nitrogen.

A significant transfer of fixed nitrogen also occurs in algal symbiosis with many higher plants. It has been demonstrated experimentally that when blue-green algae are exposed to labeled nitrogen, the nitrogen transfers quite rapidly from the algae to the leaves, internodes and roots of the plant, usually within a period of one and one-half hours (16). Thus, the algae are able to supply a high percentage of the nitrogen required for the growth of the plant.

In all of these symbiotic associations, the algae are able to grow at very low light intensities. Thus, if nitrogenase enzyme repression is eliminated in blue-green algae, and available metabolic carbon to accept the fixed nitrogen is also limited, it is not surprising that a high proportion of the nitrogen fixed is secreted extracellularly in symbiotic biological systems. As the extracellular nitrogen is generated in symbiotic systems, there is also valid evidence that the available nitrogen can be utilized by a variety of non-nitrogen-fixing plants and may be the most important source of nitrogen for the growth of these plants in their natural ecosystems.

Rebuilding Desert Soils

When building fertile soil from desert sands, it is critically important to increase the organic matter content as efficiently as possible. The addition of selected blue-green algae, which supply photosynthate derived from the atmosphere to the soil, stimulates ancillary bacterial and fungal growth by making metabolizable carbon available for their energy requirements. Thus, the blue-green algae would be of prime importance in establishing a viable soil microecology in arid lands without use of excessive energy supplies from exogenous sources.

When it is realized that many of the productive agricultural soils in the world are threatened by serious erosion and loss of vital organic matter, the hidden cost of the practice of straight commercial fertilization and continuous monoculture begins to become quite obvious. According to present-day orthodox agricultural practices, as long as the farmer supplies nutrients to the plant it is of very little consequence what happens to the organic matter in the soil. In fact, some of the more pragmatic scientists in some of the agricultural schools have assumed the posture that soil is only a mechanical device for supporting plants while farmers feed them nutrients. Unfortunately, this view overlooks the drop in agricultural productivity or marked increase in phytopreditor problems which occur when essential organic matter in the soil falls below one percent.

The biological agriculturalist, on the other hand, has demonstrated that a healthy, well-balanced soil containing adequate levels of organic matter and essential minerals will develop and grow plants in abundance utilizing less water and nutrients. It is well known that in a soil containing adequate organic matter water retention is vastly improved, and nutrient leaching due to excessive watering is reduced significantly. The reduction of organic matter in the soil not only profoundly affects phytopreditor activity, but also directly correlates with deterioration of soil tilth and fertility. Such a reduction results in an increase of soil compaction, with its lack of water absorption and resultant water run-off.

It is the business of the arid land farmer, if he wishes to survive, to function on the side of nature rather than operate against it. By treating the soil as a vibrant, living ecosystem rather than just dirt, farmers can create highly productive agricultural systems in tune with nature. If we understand the methodologies of the orient, where soil tilth and fertility have been maintained for 4,000 years through the use of biological agriculture, and we interpose our new biochemical and microbiological developments, the deserts and arid lands in the Middle East and rest of the world can be made productive agricultural areas capable of supplying the food requirements of the expanding populations.

References

1. Flohn, H. Investigation on the tropical easterly jet stream. Bonner Meteorolog Ische Abhendlungen No. 4, 1964.
2. Moreau, R.E. Vicissitudes of the African biomes in the late Pleistocene. Proceedings of the Zoological Society of London, Volume 141, 1963.
3. Jackson, J.K. Changes in the climate and vegetation of the Sudan. Sudan Notes and Records, Volume 38, 1957.
4. Raikes, R. Water, Weather and Pre-History. John Baker, London, 1967.
5. Wilton, J.I. The Roman Empire. Proceedings of the Historical Society of Bari, Volume 22, 1950.
6. Cloudsley-Thompson, J.L. and M.J. Chadwick. Life in Deserts. G.T. Foulis, London, 1964.
7. Amiran, D.H.K. Arid zone development: a reappraisal under modern technological conditions. Economic Geography, Volume 41, 1965.
8. Gustafson, C.D. et al. Drip irrigation worldwide. Proc. of the Second International Drip Irrigation Congress, University of California, San Diego, 1974.
9. Burdon, D.J. and G. Otkun. Hydrogeological control of development in Saudi Arabia. International Geological Congress, 23rd, Prague, 1968. Volume 12, Academic Press, Prague.
10. Jones, J.R. Ground water exploration and development in Libya. Water Well Journal 20:2, 1966.
11. Ibid.
12. Cooper, C. Oil drillers hit water in Sahara. Washington Post, March 15, 1969.
13. Worne, H.E. The role of soil microorganisms in seed germination and plant growth. Texas Wildflower Conference, Dallas, Texas, 1976.
14. Worne, H.E. The activity of mutant microorganisms in the biological treatment of industrial wastes. Aqua Sana Congress, Ghent, Belgium, 1972.
15. Subrahmanyan, R., L.L. Relwani, and G.B. Manna. Nitrogen enrichment of rice soils by blue-green algae. Proc. Indian Academy of Science, 1965.
16. Fogg, G.E. The production of extracellular nitrogenous substances by a blue-green algae. Proc. Royal Society of Britain 139, 1952.

Norman J. Rosenberg

10. Technologies and Strategies in Weatherproofing Crop Production

Introduction

What are the primary weather- and climate-related problems in crop production? Depending on the region and the crop involved, the limiting factor may be availability of sunlight, availability of soil water, or length of the growing season.

Production of harvestable plant products depends ultimately on the photosynthetic conversion of carbon into sugars and other more complex compounds. Photosynthesis requires energy which is provided through irradiation of the plant by sunlight.

In order for the plant to function well -- to carry out complex chemical reactions, maintain turgor and avoid overheating due to absorption of solar radiation -- a good supply of water must be available. Further, the plant must be able to extract that water with its roots and must be capable of transporting it from root to leaves and into the atmosphere at a rate determined by evaporative demand of the atmosphere.

Duration of the growing season is limited only in the boreal and temperate latitudes, but the latter are among the most productive, agriculturally, of any in the world. Occasional frosts do disturb the production of crops grown at high altitudes in the subtropical regions, when outbursts of cold air invade these regions. To grow and reach a harvestable condition, the annual crop must emerge from the soil after killing frosts end in spring, and must complete its growth cycle before the first killing frosts of autumn. Perennial plants must remain dormant during the winter and, optimally, break dormancy only after the frosts of spring are past.

Crops are distributed throughout the world on the basis of natural adaptability. Bananas do not grow in Labrador, nor does buckwheat grow in Senegal. Sustained production of a crop in any region where the normal sunlight, soil moisture and growing regime are unfavorable is not likely. But certain crops have been adapted to grow within a relatively wide range of climatic conditions. Wheat, for example, is found in the regions where temperate climates yield to the subarctic. This crop is also grown on the fringes of subtropical deserts and even within the deserts when irrigation water is applied.

Climatic Determinants of Crop Productivity

Sunlight

There is little to be done to protect or weather-proof crops from a shortage of sunlight caused by unusual spells of heavy and/or persistent cloudiness. The rates of photosynthesis, especially in plants with the C-4 photosynthetic pathway (maize, sorghum, sugar cane, millet), will decline in proportion to the shortage of light. The photosynthetic mechanism of these tropical grasses is not normally light-saturated in the temperate latitudes, even at full noontime irradiance. Plants with the C-3 photosynthetic mechanism on the other hand, generally become light-saturated at irradiances considerably lower than actually occur in the field. Hence, cloudiness may have a lesser effect upon their production. Since cloudiness actually increases the proportion of diffuse to direct beam solar radiation, the penetration of light into the plant canopies, where much of the total leaf area is shaded on sunny days, may actually increase the photosynthetic efficiency of these leaves.

Thus we see that sunlight can be in short supply. But there are circumstances, as well, when solar irradiance can be too great. Only about half of the flux density of solar radiation occurs in the photosynthetically-active waveband (400-700 nm). Of this, only one to two percent is actually involved in photosynthesis and fixed in a chemical product; the remaining energy received from the sun (all wavelengths) is absorbed or reflected by the surfaces on which it impinges (soil, water, plants). Unless evaporation, conduction or convection dissipates the heat, extreme temperatures can be reached. In the case of plant leaves, such temperatures can be lethal.

Water Supply

Water supply limits crop production in the semi-arid and

subhumid regions of the world. Many climatologists and agronomists believe that these are the regions most likely to be used in the coming decades to meet the food needs of a growing human population. The regions in question include the Great Plains of the United States and Canada, parts of Mexico, the steppes of the Soviet Union, the Pampa in Argentina, parts of northeast Brazil, large parts of Southern Africa, Sahelian Africa south of the Sahara, and large parts of India and Australia. Coastal regions in North and South America and the Mediterranean region are also semi-arid to subhumid.

Generally, these regions have abundant sunshine and wind -- both needed for growing crops -- but water is in deficit supply. In these regions on an annual basis evaporation from soil and water surfaces and transpiration (evaporation of water that has been absorbed from the soil and transported to the leaves and other evaporating plant tissue) generally exceeds the total annual precipitation. It is also true that in such regions very critical water shortages may develop in certain seasons. For example, in the Mediterranean climates, rainfall is often sufficient for crop production during winter, but the total lack of rainfall in summer limits production in that season, especially on soils of low water-holding capacity that carry over little from the rainy season. In most of Sahelian Africa, rain falls only between May and September. Crop production is virtually impossible without irrigation in the long, hot, dry season. The western Great Plains get most of their rainfall during the summer growing season. For example, Scottsbluff, Nebraska receives 70 percent of its total precipitation between May and October, but the total annual precipitation is sufficient to sustain only forage grasses or to permit production of one crop of small grains every other year.

Tactics and strategies for weather-proofing crop production in water-deficit regions are many. The development of irrigation where water supplies, soil type, land form, capital availability and the economics of energy input permit is, of course, one of the most certain of strategies. But more can be done than is done now in many regions to utilize precipitation more effectively. Precipitation must be captured and its infiltration into the soil or its storage in temporary impoundments must be facilitated. Fig. 1 shows tests of various forage species in microcatchments (runoff water captured and held in small basins) in a demonstration of runoff farming in the Negev desert of Israel. The development of crop varieties that make more efficient use of soil water is another prime strategy. This can be accomplished in many ways -- as will be explained later in this chapter. Plants

Fig. 1. Tests of various forage species in micro-catchments (runoff water captured and held in small basins) in a demonstration of runoff farming in the Negev desert of Israel.

with deep rooting habits, as an example, extract moisture from a larger volume of soil and are thus able to withstand long rainless periods.

Growing Season

Some of the world's most productive land is found in temperate regions, where cold weather limits the length of the growing season. Length of the growing season is a statistic -- the valid definition of which is debated among climatologists and agronomists. Published summaries are usually based upon the dates of last spring occurrence and first fall occurrence of 32 degrees F minimum temperature although, depending on the crop of concern, lower minima may be used. Conveniently, frost dates are normally distributed so that the probability of growing seasons of specified length is easily calculated. These numbers can be used as a guide to planting and harvesting strategies and for choosing cultivars appropriate to the length of the growing season.

Generally, the last spring and first fall frosts are radiative types involving ground-based temperature inversions. In such frosts, protection is possible by means, for example, of space heating or radiative heating, disruption of the thermal inversion by air mixing, insulating the crop with covers or foams or with a layer of ice applied as water by sprinklers or by altering longwave radiative emissivity of the crop with coatings of various kinds. The cost of protection must be weighed against the market advantage to be gained from an extended growing season in each case.

Even in subtropical regions where the "season" is normally 365 days, an occasional frost may cause severe economic loss of high-value vegetable and fruit crops. The production of an entire season may be lost or valuable perennial orchards or plantations may be taken out of production for a few years or even be totally destroyed.

Accurate predictions of the degree and duration of frost conditions are essential, but such forecasts are of little value unless protective measures are possible. The measures that are most commonly used in high-value citrus and other fruit orchards involve the disruption of thermal inversions by means of engine-powered wind machines or by heating the air under the inversion. Heating is accomplished through the combustion of various types of fuels.

Alternative means of frost protection that consume less fuel are needed. For example, it has been shown in a Texas experiment (1) that heat loss from a grapefruit grove can be diminished by lowering the thermal emissivity of the leaves with a coating of an aluminum-containing material. A plastic screen has been used to surround and cover a fruit orchard in Idaho (2). During a frost, the screen is wetted by sprinklers and the resultant ice shield protects the enclosed vegetation from the cold external environment. Within the screen, sprinklers coat the plants with water, maintaining the plants and air at temperatures near 0 degrees C. Foams have been used successfully to cover young plants during a frost night in a vegetable production area in Canada (3). Further development of such nonheating techniques will involve effective micrometeorological research. The adoption of these methods will rest on climatological studies of the frequency with which they can be successfully used. For thorough review of the principles of frost protection, see Crawford (4) and Businger (5).

Protection is possible against radiative frosts, but when outbreaks of cold air (advective) freezes occur, very little can be done except perhaps for radiative heating. Only high-value crops (fruits, vegetables, coffee, etc.) can justify the expense involved in frost protection. A more practical approach involves the careful location of crops in regions where the frost risk is economically tolerable. Knowledge of the probable duration of cold spells as well as the probable temperature minima, along with a thorough knowledge of the crop and its tolerance limits, is needed in order to properly locate crops.

Plant breeders are attending to this problem of short season or frost risk by producing cultivars of a number of species that are more tolerant of cold pressure. The mechanisms of cold tolerance and cold resistance are physiological and biochemical in origin and this chapter will not provide detailed discussion of them. The interested reader is referred to Levitt (6) for an authoritative review of this subject.

Weatherproofing Against Other Risks

Wind Damage

Mechanical destruction of crops in severe weather is another constraint to stable agricultural production. Hurricanes and cyclones do considerable damage to crops grown in certain coastal regions. This aspect of their destructiveness is rarely commented on, in view of the greater drama attached to loss of life and property.

Strong and damaging winds often reduce agricultural productivity inland as well. Cold winds in spring and fall may cause mechanical damage to the whole plant as well as freezing damage to certain tissues. Winds blowing from arid into semi-arid and subhumid regions can also cause mechanical damage. But these winds, because of their high temperature and low humidity, also impose severe moisture stress on the growing crops and cause wilting, desiccation, and the loss of potential productivity (Fig. 2). In regions where the land is not well protected by vegetation, wind erosion may occur and initiate a decline in productivity. Young, tender vegetation may be damaged or destroyed by "sand blasting" when soil is eroded by wind. For a detailed review of wind effects on plant growth, see Sturrock (7).

Properly designed windbreaks can aid greatly in stabilizing agriculture in regions where strong winds are common. The windbreak aids in uniformly distributing snow over the fields, thereby increasing the supply of soil moisture in spring. Windbreaks have a considerable impact on the crop they shelter during the growing season as well.

Considerable experimentation with tree windbreaks and with windbreaks constructed of such materials as snow fencing, plastic screens, and reed mats has shown that the climate that prevails in the sheltered area is more moderate than that in adjacent unsheltered fields. The air is slightly warmer by day and slightly cooler by night, but absolute humidity is greater by day and by night. The overall effect on the microclimate is to moderate evaporative demand and moisture stress on the sheltered plant. Since moisture stress leads to wilting, closure of the plant stomates and cessation of photosynthetic activity, the windbreak should permit the achievement of improved crop yields. Evidence from around the world showing this to be true is given in a review of "shelter-effect" by Rosenberg (8).

Despite the proven beneficial effects of windbreaks planted in the Great Plains during the drought years of the 1930s, many of them are now being removed. Changes in agricultural land use that involve larger fields and expensive irrigation systems have increased the value of the land to the point where farmers begrudge its occupation by tree windbreaks. Windbreaks may interfere with the mechanical operation of the large center-pivot sprinkling systems that are revolutionizing irrigation in the Great Plains region (9).

There is urgent need for windbreak designs that are

Fig. 2. An artificial windbreak at Gilat, Israel for the study of shelter effect on evaporation from bare soil.

compatible with current and foreseeable agricultural systems in windswept regions.

Disease and Insects

Three conditions must be met before disease or insects can infest a crop: the disease organism or insect must be present, the host plant must be susceptible to attack by the parasite, and the environment must be conducive to activity and reproduction by the attacking organism. Literally thousands of diseases and insects that attack plants have been identified. There is a vast literature categorizing plant species and varietal susceptibility to diseases and insects. Unfortunately, the specific environmental conditions conducive to the proliferation and attack by specific parasites are known in only a few cases.

Today the control of disease and insects is accomplished in agriculture largely by the application of chemical sprays and dusts. Increasingly sensitive to the introduction of chemical agents into the food we eat and the air we breathe, our society is becoming more critical of this approach. Methods other than chemical control or those involving limited use of chemicals for control of disease organisms and insects are now being developed.

An appropriate strategy for improved chemical control of pests may involve the following steps:

1. Define the environmental conditions conducive to disease or insect outbreaks, e.g. temperature, humidity, leaf wetness, windiness, illumination.

2. On the basis of weather records, determine the probabilities that these environmental conditions will prevail at any specific time and for any specified duration.

3. On the basis of up-to-date information, prepare weather forecasts indicating the probability of occurrence of conditions conducive to the disease. Only then would actual spray operations be considered.

Such a combination of biology, climatology, and meteorological forecasting has been used to develop effective systems for control of the late-blight disease of potatoes in Ireland. These systems are used in other potato-growing nations, as well.

Agrometeorologists may help in new ways to develop strategies other than chemical control of plant pests. The white-mold disease of beans provides an example. This disease flourishes in humid regions when the weather is cool and damp for protracted periods. The disease has also become important in semi-arid regions where irrigation of bush varieties creates a conducive microclimate. New designs of the crop-canopy architecture, which can be introduced by plant breeding in conjunction with altered irrigation schedules, are now under study (10). These adaptations may permit the maintenance of a microclimate less hospitable to the white-mold disease.

There has been other evidence of progress in applying meteorology to the control of plant disease. The spread of southern leaf blight of corn, for example, which threatened much of the 1970 crop in the corn belt, has been well modelled (11). An alert system was devised using a computer model in which observations on duration of leaf wetness, air temperature and humidity were obtained from zones in which the crop is susceptible to the disease. Models of this kind can be developed for other diseases, given adequate knowledge of the biological relationships and some understanding of the mechanisms by which spores or other vectors of disease are transported from zone to zone.

The Great Plains as a Case Study

Drought is experienced in every climate except in the deserts of the world. This seeming paradox is easily explained. Drought is (I choose from among the hundreds of definitions one that is pertinent to this discussion), "a climatic excursion involving a shortage of precipitation sufficient to adversely affect crop production or range productivity." Experience indicates that such droughts, if sufficiently protracted, affect the economy and society as well as the politics of the region in which they occur. Deserts, with little or no rainfall, cannot suffer the effects of any unusual water shortages.

The Great Plains of the United States and Canada provide ample evidence of a wide range of skills gained and strategies attempted in coping with drought. The region is one in which protracted droughts have occurred with some regularity since meteorological record-keeping began. There are also good surrogates for meteorological observations, such as tree-ring records, giving evidence that droughts occurred in pre-settlement times.

In economic terms, drought losses can be considerable in the Great Plains. For example, dryland corn yields were reduced drastically in 1974 and 1976 during what turned out to be short, but relatively severe spells of drought. In 1975, a good year from the point of view of precipitation amount and distribution, corn yield was 48 bushels/acre; it was 28 bushels/acre and 40 bushels/acre in 1976 and 1977, respectively. Perhaps more significant is the fact that the number of acres of dryland corn harvested declined by 350,000 acres in Nebraska between 1974 and 1976. In 1974 loss of production of soybeans, corn, sorghum, wheat and hay amounted to $1 billion in Nebraska alone (12).

In 1979 an analysis of the impacts of drought on the Great Plains and of the research needed to undergird the development of management strategies was undertaken with support of the National Science Foundation, Division of Problem Focused Research. A Workshop Research in Great Plains Drought Management Strategies was held for this purpose in Lincoln, Nebraska on March 26-28, 1979. Task Groups were assigned to study four aspects of the drought problem: 1) Technology; 2) Economics; 3) Political Considerations and Decision-Making; 4) Social and Information Problems. The full proceedings of this workshop were in press (Rosenberg, ed., 1980) when this was written. A synopsis of the findings of the Technology Task Group is given (13), since this group dealt most closely with ways of weather-proofing crops under drought conditions. The Task Group categorized the possible strategies as short-term or long-term measures.

Short-Term Measures

In times of actual drought, particularly if the season is advanced, only few technological options are available to dryland farmers. Reduction of plant populations -- as by removing alternate rows -- can be done to reduce plant competition for available water. Fertilizer applications can be withheld to avoid further stressing the plants by increasing osmotic potential of the soil solution. Projected yield losses can be reduced by cutting crops green for silage rather than waiting to harvest a very poor yield of grain.

Irrigation farmers have other options. Drought may encourage drilling of more wells, or existing wells may be pumped for more hours per week. These tactics may actually be self-defeating in the long run, especially in areas where irrigation depends on fossil water. This is true since accelerated depletion of fossil water supplies means that coping with the next drought will be still more difficult.

Irrigators have other, perhaps more logical options. One is to irrigate more land with less water applied to each unit of land. In times of drought, crop water demand in irrigated fields is greatly increased by the advection of sensible heat from dry surrounding fields and from dry surrounding regions (14, 15). Irrigation systems are generally not designed to cope with prolonged periods of such extreme evapotranspiration rates as occur during drought. By limiting irrigation to critical periods in the crop life cycle (tassel initiation in corn for example) the greatest yield can be achieved per unit of water applied.

Gravity irrigation systems are easiest to manipulate in this way. The farmer can furrow irrigate only in alternate rows, for example. Fixed sprinkler systems are more difficult to adapt, but even the massive center-pivot irrigation systems are designed to be moved from one quarter section of land to another. Older sprinkler systems such as the "skid-tow" were designed for easy mobility. Thus the acquisition of skid-tow pipe for emergency use may make good economic sense. New, more mobil irrigation systems that can be quickly adapted to irrigate large areas with relatively small quantities of water are certainly needed.

<u>Long-Term Measures</u>

Technological measures to mitigate drought must accomplish at least one of two purposes: increase water supply; reduce water demand. Long-term measures that accomplish these ends are likely to be helpful in years when precipitation is good, as well as during drought years. Most are proven practices representing good land management. A few are less well established, but their potential justifies a serious research effort.

<u>Minimum tillage</u>. Methods have been developed to reduce the number of tillage operations needed in crop production. Certain plowing, harrowing and cultivation operations can be eliminated for many crops -- particularly where chemical herbicides are effective in weed control. The benefits of minimum tillage in reducing soil erosion by wind and water have been demonstrated (16) as has the increased soil moisture availability in times of drought (17). About 20 percent of U.S. crop production is on minimum or no-tilled land at this time. Some unsolved problems of minimum tillage include uneven seed germination, low soil temperature in spring, possible disease and insect outbreaks and possible undesirable environmental effects due to reliance on chemical herbicides. Nonetheless, the potential of minimum-tillage methods for improving soil moisture conditions, minimizing losses of

top-soil and reducing the energy costs in crop production indicates that adoption should increase in coming years.

Snow management. In the Canadian Prairie provinces, the northern Great Plains and into Nebraska and Kansas a significant portion of the annual precipitation occurs in the form of snow. Unless controlled, snow either blows off or runs off over frozen ground as it thaws. Thus, snow often contributes little to the reservoir of soil moisture available to the crop in the spring. Snow is best controlled by reducing windspeed near the surface. This can be accomplished with constructed barriers, tree windbreaks, sown windbreaks or annual or perennial crops (18) or by leaving stubble of the previous crop standing in the field, as shown in Fig. 3.

Barriers cause some problems. Efficiency of tillage operations is decreased to a degree where barriers such as trees create traffic obstacles. Windbreaks may harbor insects that attack the sheltered crop, but they may also, of course, harbor beneficial creatures. The incidence of fungal diseases is thought to increase where the windbreak creates a more humid environment that can favor such disease. Nonetheless, the benefits of wind barriers, particularly in snow management, are important. In the semi-arid regions where grains are grown, even small increments of water lead to very significant yield increases.

Irrigation scheduling. Irrigation provides an important degree of stability to the food-producing industry of the United States. More than 60 million acres of land are now irrigated in the United States. Of these, 51 million are in the 17 western states. Growth of irrigation has been rapid in the recent past. In the Great Plains region, irrigation has risen at an exponential rate since the 1950s. In Nebraska some 6.8 to 7.0 million acres are now irrigated.

The greatest increase in irrigation in recent years in the Plains region has been with sprinkling systems -- primarily the center-pivot system, which permits irrigation of from 135 to 152 acres in each quarter section (160 acres) land to be irrigated. Center pivots are expensive to install -- upward of $60,000 to $75,000 per system (including well preparation and pump installation) in 1979. Most of these systems operate at high pressure (60 to 80 pounds per square inch) and are powered by electric motors or by engines that utilize diesel fuel, liquefied petroleum gas or natural gas.

At 1979 prices, a single irrigation cycle costs between $4 and $6 per acre. In all, during an average year in Nebraska, for example, fuel consumption for a single

Fig. 3. Stubble mulch farming in Western Nebraska. Wheat straw is left standing in the field after harvest in order to help spread snow and reduce wind erosion until the next crop is planted.

center-pivot system may range from 4,500 to 10,000 gallons of diesel equivalent. Since existing pivot systems in the United States now number over 50,000 and irrigate more than seven million acres, the fuel consumption involved approaches half a billion gallons diesel equivalent per year.

Excessive irrigation -- more frequent and more intense application of water than is needed -- wastes fuel, degrades soil quality, depletes groundwater supplies more rapidly than necessary (thus increasing pumplifts and further increasing fuel requirements), and may degrade groundwater quality. Excessive irrigation can actually reduce crop yield. Most farmers tend to apply excess water; many overirrigate by 50 to 100 percent.

Irrigation can be scheduled on a field-by-field basis if certain knowledge is available:

a. Soil type, moisture-holding capacity, and antecedent soil moisture content;

b. Crop type, stage of development, and water requirement at each stage;

c. Irrigation system characteristics and efficiency;

d. Cumulative evaporation since the last irrigation or rainfall event and probable evapotranspiration and precipitation in the forthcoming week to 10-day period;

e. Forecast information made available on a timely basis as the irrigation approaches.

Since two to four days are normally required for irrigation of a 160-acre field, forecasts must provide for an appropriate lead time or rapid updating.

Climatological and weather information must be entered daily or, minimally, twice weekly into the decision matrix in order to facilitate scheduling of each irrigated field. Scheduling services are currently being provided by many small companies scattered through irrigation regions. These services often rely on rough estimations of evapotranspiration or on observation of a few soil moisture samples. A few larger companies make use of computer technology to monitor conditions in large numbers of fields. The AGNET interactive computer-information system, in use in Nebraska and adjacent states, is already used to provide scheduling advice to a large number of irrigators (19). But here, too, climate and weather information is not received in a manner permitting maximal efficiency and accuracy. Thus, there is an obvious need for better coordination of weather data, climate probability information, evapotranspiration models and information dissemination systems to improve irrigation scheduling and make it a readily accesible tool for water management at all times, but especially in drought.

Microclimate modification. Windbreaks, as discussed above, alter the microclimate of plants in their lee. In general, air temperature is increased by day and lowered by night, but absolute and relative humidity are increased at all times. Moisture stress on the plant is minimized and stomata in the leaves of sheltered plants often remain open while turgor loss in unsheltered plants causes stomatal closure, and hence, a loss of photosynthetic opportunity. The windbreak is one of the better-known tools of microclimate modification used in semi-arid and subhumid regions to improve crop performance. Other methods are also available. These include mulching the soil with crop residues or transported materials such as plastic sheeting, gravel, coal cinders, etc. Mulching suppresses direct evaporation from the soil surface and thus preserves water for later use by the crop. Heavy mulches of straw from small grain harvests appear to be effective and economical in saving water for the next crop. Some of the imported materials, on the other hand, may be too expensive for use with any but specialized high-value crops.

Antitranspirant chemicals have been applied to growing plants to reduce their water use. Certain of these chemicals (e.g. phenyl mercuric acetate and Atrazine) function by inducing stomatal closure. The stomatal resistance to diffusion is of considerably greater relative importance in the transport of water vapor out of the plant leaf than it is in the transport of carbon dioxide into the plant leaf. Hence, induction of stomatal closure should have a positive influence on water use efficiency -- the ratio of dry matter produced in photosynthesis to water consumed in transpiration. In practice, for environmental and other reasons, chemical antitranspirants have not been widely adopted.

Long-chain alcohols and waxes have also been used as antitranspirant materials. These materials, when applied to the leaf, create a physical barrier to water vapor diffusion. These barriers are also more permeable to carbon dioxide diffusion -- at least in theory. Acceptance of these materials has also not been widespread.

Reflectant materials may also be used to suppress direct

evaporation from the soil surface or transpiration from the plant, as shown in Fig. 4. Soil surface albedo can be increased from around 10 percent in dark soils to perhaps 50 percent by application of kaolin, white wash, diatomaceous earth or other appropriate material. This increase in albedo means, simply, that a greater portion of the incident solar radiation is reflected back to space. That radiation were it absorbed would warm the absorbing surface and contribute to evaporation of water from it.

Reflectants have also been applied directly to crop surfaces. In a series of studies at Mead, Nebraska (20, 21, 22, 23), kaolinite clay or diatomaceous earth (Celite) were sprayed in a slurry mixture onto a soybean crop. The photosynthetic mechanism of soybeans is light-saturated at about 700 Wm^2, which is about half the radiant flux density of sunlight at noon in this locality. Thus, the solar radiation unneeded in photosynthesis actually increases the heat load on the plants, and this heat is dissipated by transpiration.

In the experiments described above, a 15 percent savings in water was accomplished with no concomitant reduction in photosynthesis. These experiments have provided one practical (if expensive) technique for reducing water use in times of drought. Perhaps, more importantly, they have provided a clue to plant breeders on one way (altered albedo) in which plants can be redesigned to use less water -- a means of weatherproofing crops.

Alternative crops. One way to face a problem may be to avoid it. Thought is being given to introduction to the Great Plains regions of new crops that are less sensitive to moisture stress than those currently grown. Possible new food crops for use in dryland farming systems or with limited irrigation include pearl millet, amaranth and guar. Specialty crops considered for introduction include guayule as a source of latex and forage sorghum for biomass conversion. Kochia and fourwing saltbush (*Atriplex canescens*) are potentially useful new plants for Great Plains rangeland.

Breeding, cropping systems development and marketing research will be needed before these crops can be introduced.

Improved cultural practices. Tillage, fertilization, strip cropping, skip-row planting and crop rotation are examples of cultural practices that can be used to minimize the impact of drought by maximizing crop production while achieving water and soil conservation. The application of these practices to specific crops, climates and economic circumstances requires practical agronomic research.

Fig. 4. Soybeans at Mead, Nebraska, coated with Celite to increase reflection of unneeded solar radiation and consequently reduce crop water use.

Water harvesting. Water harvesting is the capture of runoff water which spreads over depressional areas in fields or the floodplains of streams. The practice has ancient roots. Evenari et al. (24) have described reconstruction of ancient Nabataen settlements in Israel's Negev desert. Water spreading has been limited mostly to a few high-value crops grown in the Southern Plains region, but the potential exists for greater use of microcatchments in low rainfall areas.

Crop selection for drought tolerance. Crop cultivars that perform well under drought stress are needed to stabilize yield performance of the major crops grown in the Great Plains region. Drought tolerance has not been a major objective of plant breeding in the region. Instead, the need to respond to disease and pest problems and a general aim of breeding for high yield (under optimum conditions) have taken precedence. Because of the adaptive nature of most pests, the need to breed for resistance against them is almost perpetual.

It is clear that two approaches must be used in breeding plants for drought tolerance:

1. Continued conventional plant breeding research is needed to produce new cultivars with superior drought tolerance.

2. Basic research is needed on the morphological, physiological and biochemical characteristics that provide drought tolerance, with the aim of using these characteristics as tools for screening new lines or of incorporating them into new cultivars.

Micrometeorologists, using basic theory and computer modelling, are able to define these plant morphological or "architectural" characteristics that best fit the plant to its microenvironment. The influence of such factors as plant shape, leaf shape, leaf orientation with respect to the sun, total leaf area and leaf area density, leaf roughness, leaf pubescence, plant height, stiffness or flexibility in wind, among others can be modelled and "blue prints" provided to plant breeders.

Until now such exercises have been essentially academic. However, geneticists are now able to produce "isogenic" cultivars in a few of the major plant species. Isogenes are plants identical in every way except for a single character controlled by a single identifiable gene. Ferguson et al. (25), for example, report the existence of "golden" and normal isogenes of barley. The golden cultivar is light green in color and reflects more sunlight.

Isogenes of the soybean cultivars Harosoy and Clark have been bred to differ in plant height (because of node number or node length), leaf shape (normal oblate or elongated), leaf roughness (normally smooth or crinkly), leaflet number (normal trifoliate, 5-foliate or 7-foliate) and pubescence (normal, high pubescent or glabrous) (26).

This wide range of architectural options in lines that are otherwise genetically identical provides a major scientific opportunity to test the idea of tailoring the plant to its physical environment. Adaptations most appropriate for drought conditions may be shortened stature, increased reflectance and increased leaf roughness.

Conclusions

Weatherproofing crops may be approached from the conservative viewpoint that addresses only adversity -- what to do in cases of drought, hail, frost, etc. Or, we may consider the factors that limit crops from achieving their inherent productive potential and then systematically remove these obstacles.

Sunlight, water and a viable temperature regime are required if crops are to germinate, grow and yield their product. When any one of these factors is lacking for a significant period of time the final yield of the crop is diminished. Thus, the first and most efficient weatherproofing technique is proper selection of the land and climate in which the crop can be grown with minimal weather risk. In a sense, the current worldwide distribution of crops represents an historic process of trial and error by all the farmers who have ever lived. Of course, the distribution is affected by manifold cultural and economic factors, and the optimum distribution of crops is yet to be achieved.

Most people think of weatherproofing in terms of crop damage that occurs under episodic weather events -- windstorm, killing frost, hailstorm. There are effective protective techniques in some cases: windbreaks, orchard heaters, wind machines, insulating covers, etc. The utility of these techniques is determined by the value of the product that can be saved and the cost of the protective measure.

To focus this subject of weatherproofing crops, I have concentrated the preceding pages on currently available and prospective technologies that can be used in drought-prone regions.

The increased capture of precipitation and storage of the captured water in impoundments or preferably in the soil, is the first strategy considered. Snow management affords a good opportunity in the northern Plains, especially, to increase storage of water in amounts that can significantly increase crop yields. Winter harvest methods that capture and spread runoff water on lower-lying lands are practiced in many semi-arid areas. Minimum tillage reduces the number of field operations that expose soil to evaporation and leaves a residue to plant material on the surface to further reduce evaporation. Stubble mulching in grain fields improves the capture and retention of snow and tends to reduce direct evaporation from the soil surface. Straw mulches and mulches of other materials tend to suppress evaporation, leaving more water for subsequent crops. Windbreaks, similarly, can reduce direct evaporation from the soil.

Irrigation is an obvious drought-proofing technique. But strategic irrigation -- using water on as wide an area as possible, especially at critical stages of plant growth -- can be much more effective in drought than irrigating smaller areas intensively. Irrigation scheduling -- determination of the most appropriate time and amount of irrigation water to apply -- is an important tool in stretching available water supplies. Calculations are based on computation of antecedent soil water status, current and projected evapotranspiration rate, stage of crop growth, forecasted near-term weather conditions, climatological probability of longer-term water requirement, and irrigation system capabilities and constraints.

Optimal use of soil-available water and increased water-use efficiency can be accomplished by another set of weatherproofing tactics. Microclimate modification by means of windbreaks, antitranspirant materials applied to the plant, reflectants applied to soil or plant, and other techniques can be used to reduce transpiration, increase photosynthesis, or accomplish both at the same time. Improved cultural practices involving tillage, fertilization, strip cropping and skip-row planting can also reduce the stress on plants' growing in drought conditions by reducing plant competition for remaining water.

Alterations to the plant itself may provide the best of future strategies to weatherproof crops. Theory suggests that optimal architectures can be designed to minimize the physical stress imposed on plants by the environment. Advances in plant breeding may make it literally possible to construct plants having optimal architectures.

Another serious option is the introduction of new crops to drought-prone regions. These are crops that derive from the arid zones and already possess architectural features and physiological mechanisms that provide some degree of drought tolerance. Adaptive research is needed and markets for such crops must be developed before a major conversion can be expected.

References

1. Hales, T.A. Alteration of nocturnal radiation balance of "red blush" grapefruit trees by application of aluminum powder suspensions to upper leaves. Agric. Meteorol. 13: 59-67, 1974.
2. Cary, J.W. An energy conserving system for orchard cold protection. Agric. Meteorol. 13:339-348, 1974.
3. Desjardins, R.L. and D. Siminovitch. Microclimate study of the effectiveness of foam as protection against frost. Agric. Meteorol. 5:291-296, 1968.
4. Crawford, T.V. Protection from the cold. Frost protection with wind machines and heaters. Agric. Meteorol. 6: 81-87. Amer. Meteorol. Soc., Boston, 1965.
5. Businger, J.A. Protection from the cold. Frost protection with irrigation. Agricultural Meteorology, P.E. Wagonner, ed. Chapter 4, Section 1. Meteor. Mongr. 6: 74-80, Amer. Meteorol. Soc., Boston, 1965.
6. Levitt, J. Responses of Plants to Environmental Stress. Academic Press, New York, 1972.
7. Sturrock, J.W. The control of wind in crop production. Progress in Biometeorology, L.P. Smith, ed., Volume 1, Swets and Zeitlinger, Amsterdam, 1975.
8. Rosenberg, N.J. Windbreaks for reducing moisture stress. Amer. Soc. Agr. Engin. Monograph on Modification of the Aerial Environment of Plants, B.J. Barfield and J.F. Gerber, eds., 1979.
9. Goldsmith, L. Action needed to discourage removal of trees that shelter cropland in the Great Plains. Proc. Symp. on Shelterbelts on the Great Plains. R.W. Tinus, ed., Great Plains Agr. Council Publ. Nol 78, 1976.
10. Blad, B.L., J.R. Steadman and A. Weiss. Canopy structure and irrigation influence white mold disease and microclimate of dry edible beans. Phytopathology 68:1431-1437, 1978.
11. Wagonner, P.E., J.G. Horsfall and R.J. Lukens. A simulator of Southern Corn Leaf Blight. Conn. Agr. Expt. Sta. Bull. 729, 1972.

12. Nebraska drought loss is calculated at $1 billion. Lincoln Star, March 11, 1975.
13. The Task Group membership was as follows: H.E. Dregne, Texas Tech. U., chairman; Richard A. Warrick, Clark U. and Gordon McKay, Environment Canada, reporters; N.J. Rosenberg, U. Nebraska, coordinator. Other members of the Task Group were: R.D. Burman, U. Wyoming; O.C. Burnside, U. Nebraska; A.H. Ferguson, U. Montona; C.O. Gardner, U. Nebraska; A. Guitron, Government of Mexico; A.T. Harrison, U. Nebraska; J. Heilman, South Dakota State U.; H.S. Jacobs, Kansas State U.; L. Jump, International Harvester Co.; W. Willis, USDA/SEA/AR, Fort Collins, Colorado.
14. Brakke, T.W., S.B. Verma and N.J. Rosenberg. Local and regional components of sensible heat advection. J. Appl. Meteorol. 17:955-963, 1978.
15. Rosenberg, N.J. and S.B. Verma. Extreme evapotranspiration by irrigated alfalfa: a consequence of the 1976 midwestern drought. J. Appl. Meteorol. 17:934-941, 1978.
16. Greb, B.W. Reducing drought effects on croplands in the west-central Great Plains. U.S.D.A. Info. Bull. No. 420, 1979.
17. Wittmuss, H.D. and A. Yazar. Conservation tillage strategies for increasing crop production with limited water. Agron. Abstracts. Amer. Soc. Agron. 1977 Annual Meeting.
18. Rosenberg, N.J. Sown windbreaks. SPAN 20:12-14, 1977.
19. Fischbach, P.E. and T.L. Thompson. Irrigation scheduling. Technology transfer program using AGNET computer system and other tools. Presented to American and Canadian Socs. Agr. Engin. Summer Meeting, Winnepeg, Canada, June 1979.
20. Doraiswamy, P.C. and N.J. Rosenberg. Reflectant induced modification of soybean (Glycine max. L.) canopy radiation balance. I. Preliminary test with a kaolinite reflectant. Agron. J. 66:224-228, 1974.
21. Baradas, M.W., B.L. Blad, and N.J. Rosenberg. Reflectant induced modification of soybean (Glycine max L.) canopy radiation balance. IV. Leaf and canopy temperature. Agron. J. 68:843-848, 1976.
22. Baradas, M.W., B.L. Blad and N.J. Rosenberg. Reflectant induced modification of soybean (Glycine max L.) canopy radiation balance. V. Longwave radiation balance. Agron. J. 68:848-852, 1976.
23. Lemeur, R. and N.J. Rosenberg. Reflectant-induced modification of soybean (Glycine max L.) canopy radiation balance. II. A quantitative and qualitative analysis of radiation reflected by a green soybean canopy. Agron. J. 67:301-306, 1975.
24. Evenari, M., L. Shanon, and N.H. Tadmor. Runoff farming in the desert. I. Experimental layout. Agron. J. 60: 29-32, 1968.

25. Ferguson, H., C.S. Cooper, J.H. Brown and R.J. Eslick. Effect of leaf color, chlorophyll concentration and temperature on photosynthetic rates of isogenic barley lines. Agron. J. 64:671-673, 1972.
26. Sprecht, E. and J.H. Williams. University of Nebraska, Department of Agronomy, personal communications, 1978-1979.

Carl N. Hodges

11. New Options for Climate-Defensive Food Production

The tremendous impact of climate on food production can be moderated -- and in some cases virtually eliminated -- by applications of environmental control. Such beneficial manipulations range from the simple to the complex; the extent to which the technology is applied is usually determined less by the limits of knowledge than by the availability of resources and by cost-effectiveness.

Sunlight, for example, is the ultimate energy source for all food production. Plants use it to convert water and carbon dioxide into carbohydrates. Yet the major agricultural areas of the world are not the sunniest areas of the world. This is because historically a balance had to be struck between available sunlight and available rainfall, which is still agriculture's primary source of water. So the earliest and most significant type of environmental control is irrigation; by supplying water artificially to plants, it became possible to develop agriculture in areas of maximum sunlight. Such regions have become the most productive on the globe; sufficient water is available for plant growth, but clouds do not shade the land as the water is delivered.

This primary type of climate control has severe limiting constraints, however. We have already developed most of the world's potential for irrigation; we have done most of what we can where sunlight, fresh water and arable land are all available and can be joined realistically. And, in the sunny, arid regions which become so productive with the application of water -- i.e. the deserts -- we tend to load the fields with residual salts.

A desert, by simplest definition, is a place that gets less water than it loses by evaporation. About 35 percent of all the land on earth -- 18 million square miles -- is classified as desert, including a third of the North American continent. This does not mean that all deserts are hot and glaring with sunlight -- there is a big one in Asia that is frozen solid a good part of the year -- or utterly without water and uninhabitable. It does mean that they get little rain and have high rates of evaporation. The latter distinction is important because it causes us to manufacture more desert every year. We do this with the best of intentions -- we are trying to grow more food on more irrigated land -- but the simple desert phenomenon of evaporation is getting ahead of us. About a half million acres of crop land are estimated lost each year in this manner.

The problem is not so much that a great deal of irrigation water must be used, although this in itself is critical in the desert, where there is only so much water available. Sometimes the water table drops so far one can't afford to pump it up anymore. But the problem we address here is that when the water evaporates it leaves its salts behind in the soil. Year by year the land gets saltier and the crop yield gets poorer until finally practically nothing will grow. Conditions deteriorate quicker if the water is brackish, and desert water often is. The land is abandoned eventually, more unusable than ever before.

The continuous loss of arid land to salinization is nothing new. Whole ancient civilizations are believed to have wiped themselves out this way, helping create the formidable deserts of the Middle East and North Africa in the process. But the phenomenon is also dangerously contemporary. The state of California grows a large part of most American foodstuffs, yet the build-up of salt in the soil is now said to affect half of its crop lands to one extent or another. The problem is common to much of the southwest United States and northwest Mexico. The two countries share the same geographical deserts -- the Mojave, the Sonoran and the Chihuahuan -- and like those in the United States, the Mexicans haven't been at all sure what to do about salinity.

Is Fresh Water the Limiting Factor?

It is disappointing that on this lush planet, with a full two-thirds of its surface covered by water, there is so little usable fresh water available. Almost all of it is locked up in the oceans, the ice caps and the glaciers. Less than one-half of one percent of the world's water is theoretically available for crop irrigation and all other human needs. Much of it, in fact, society itself has rendered at least temporarily unusable, especially in the developed world. Yet there is an immensity of salty water confronting us, not only in the oceans, but also in the saline ground water which underlies much of the land. Two-thirds of the land mass of the United States lies above saline water of more than 1,000 parts per million (ppm). But even more compelling is the presence of 20,000 linear miles of desert seacoast in the world, backed by unknown millions of acres of empty land, drenched in the highest levels of solar radiation. Consider the coasts of North Africa, the Arabian peninsula, much of the Indian subcontinent, the Atacama of South America, the western coast of Australia, and the more than 2,000 miles of desert seacoast on both sides of Mexico's Gulf of California.

At the same time, consider also that those of us who have grown up in the desert have always been fascinated by the challenge of "making the desert bloom." This dream is exaggerated when one stands on a desert seacoast and looks out at all of that blue water offshore. If only we could remove the salt from the seawater and use it to irrigate the land. The University of Arizona's Environmental Research Laboratory (which the author directs) made an effort in this direction in the 1960s. We developed a solar-powered desalting plant that performed as designed and was probably about 20 years ahead of its time. Despite its technical success, however, this solar plant suffered from the same malady that still afflicts a great many solar applications today: it was just too expensive to consider using such product water, even under the most optimistic projections, for conventional open-field agriculture.

The increasingly prohibitive cost factors in desalinized seawater irrigation leave us with two choices: 1) We can try to reduce the water requirements of conventional food crops -- which was our laboratory's first approach to the problem, or 2) We can try to do something with seawater itself, something other investigators have tried for years, but which we are now attempting from several points of view.

First: CEA to Conserve Water

Our laboratory's first approach, from the late 1960s through the early '70s, led to sophsticated applications of Controlled-Environment Agriculture (CEA) in the deserts of the United States and the Middle East. Rather elaborate greenhouses are used to encapsulate an artificial environment. In there, air is circulated through a spray of seawater to cool and humidify it, producing a rainforest environment, but with high levels of sunlight. Vegetable production is prodigious, and little fresh water is necessary, as the humid atmosphere reduces plant evapotranspiration to a fraction of what it would be in the open desert. The net

Fig. 1. A two-hectare controlled-environment agriculture facility located on Sadiyat Island, Abu Dhabi, designed by the Environmental Research Laboratory.

Fig. 2. Superior Farming Company, Tucson. A ten-acre controlled-environment agriculture commercial facility designed by the Environmental Research Laboratory.

effect is climate control, as well as water conservation. So little fresh water is needed relative to the high levels of food production, that all other things being equal, even very expensive water from a desalting plant can be economically feasible. There are several large-scale applications of this technology producing special high-value crops in both hemispheres. Fig. 1 and Fig. 2 show CEA installations, designed by the Environmental Research Laboratory (ERL), at Abu Dhabi and near Tucson, Arizona.

The major constraint in CEA, however, is again a matter of economics. A total CEA system is capital-, energy- and labor-intensive; it may not use much fresh water, but it uses a great deal of money. In the United States, most commercial CEA greenhouses are marginal at best, and only those installations which produce for a very specialized market or for a captive market or for both make very much sense. (CEA systems are more common, and presumably, more profitable, in parts of Europe and Japan.) We hope that innovative designs and new materials will reduce future structural and energy costs, and are still working to that end.

The other approach -- using seawater directly for food production -- is one of the oldest and most cherished hopes of mankind. There are a number of options to consider.

Seawater Farming

Perhaps the most obvious of all, the growing of food using seawater instead of fresh water is in many ways extraordinarily difficult to utilize in modern production systems. Since conventional farm livestock cannot use highly salty water, we can fasten upon those economically desirable aquatic animals, particularly marine species, which can. And as it is increasingly inefficient to hunt fish on the open ocean (wasteful of energy and at the mercy of weather or unpredictable sea) we can bring the ocean onshore and there domesticate crops of sea animals -- that is, engage in mariculture, or seawater aquaculture.

It may seem a little puzzling that the developed world, with all of its technology, has seen so few successful applications of aquaculture, an art that has long provided subsistence farming in parts of Asia and Africa. It hasn't been for lack of trying. Particularly in the decade from the mid-1960s to the mid-1970s, there was a national preoccupation with "oceanography" and "aquaculture" in the United States, and people were led to believe that large-scale farming and mining of the sea were going to happen immediately. What went wrong? It seemed to involve serenely simple technology. Everybody understood how a fish pond operates, and aquaculture essentially seemed a fish pond made bigger.

What went wrong was that we "pioneering" aquaculturists really didn't know what we were doing. We knew so little about the animals of the sea, and we still know very little. We simply did not understand that fish farming involves highly complex biological systems, and in an unfamiliar and corrosive medium. To some extent we recognized our ignorance, but our method of trying to learn was groping and ill-coordinated. Instead of a strong, centralized, multidisciplined NASA-type assault on the problem, we settled for a scattered burst of unrelated, small-scale individualistic little research projects which produced hundreds of graduate degrees, tons of paperwork, and very few pounds of fish. At the same time, a handful of visionary and articulate entrepreneurs persuaded a succession of equally enthusiastic investors to initiate commercial aquaculture ventures. With few exceptions, these were mostly premature disasters. Altogether it was a rather sorry little history and created widespread disillusionment. It was comparable to some of the current disappointments in the field of solar energy; we believed because we wanted to, and we were led to expect too much too soon.

An Integrated Approach

It was in this context, in 1972, that the Environmental Research Laboratory began work in an approach combining climate control with the use of saline water in growing seafood. We did not expect the development of high technology aquaculture to be easy, and it wasn't. It took seven years and seven million dollars to develop a CEA system for the culture of marine shrimp. We deliberately selected such a high-value, luxury-food species as the initial cultured product, because there is a worldwide demand which far exceeds worldwide supply. We knew that the most efficient way for a university to disseminate a new food technology is to invent something that is clearly profitable; the dissemination then takes care of itself. We also chose to encapsulate the artificial aquatic environment, not for cooling and humidification as in the case of CEA for plant production, but rather to control light as well as to help elevate temperatures and protect a valuable crop of small animals from sea birds and other predators which so frequently devastate open-pond aquaculture.

In a large-scale project funded by the Coca-Cola Company and the F.H. Prince Company, and working with the University of Sonora, we took a species of commercial shrimp from the

Gulf of California, brought it onshore at the Mexican research station shared by our two universities, and domesticated it. In our controlled-environment systems, the shrimp breed and reproduce themselves. Their offspring are grown by the millions to harvest size on a variety of artificial diets and at high-stocking densities undreamed of in conventional pond-based aquaculture. Instead of an objective of a ton of animals per acre per year, we can grow more than 50 tons per acre-year, and of a higher quality to bring a higher price. This is more shrimp from one acre than four local shrimp boats catch in a season, and with vastly reduced inputs of capital and energy. Our research sponsors are now beginning to commercialize this shrimp culture system in large-scale applications. Fig. 3 shows the installation at Puerto Penasco, Mexico.

Growing Salt-Tolerant Crops

The other options toward "biosalinity" production deal with plant agriculture. Certainly plant science research in this direction has not been inactive, but the work has been small in scale and spread out over a longer period of time. It also in some ways appears more complex, probably because so little is known about biosaline plant production. There are essentially three approaches toward using saline water to grow plants, the first and most obvious of which is to cultivate marine algae or seaweed. So far in the United States, culturing algae on shore has proven mostly unsuitable and unattractive for human consumption, economically impractical for livestock, and marginal at best for selected chemical extraction. Chances are this situation will improve, since there appears to be no particular economic or technical barrier involved. It is more a matter of discovering and/or developing better kinds of algae for this purpose.

The second approach is to find or develop strains of conventional land-based food crops which can tolerate higher levels of salt. This is a goal as old as agriculture and over the years a great many scientists have gone after it. What scientific literature exists on biosalinity -- and there is not a great deal of it -- mostly describes these scattered efforts. The work is still going on, with a few hints and promises, but with no real breakthroughs. Salt slows the growth rates of conventional crops, reduces their quality as well as their yield, and kills them at saline levels only a fraction that of seawater.

The third approach is newer. It is to explore those strange and neglected plants which have evolved naturally in highly saline soil -- halophytes, named from the Greek words

Fig. 3. Foreground: U. of Sonora Research Station in Puerto Penasco, Sonora, Mexico, where cooperative research by the U. Sonora and the U. Arizona has continued for 17 years. Facilities for seawater aquaculture are to the left.

for salt and plants -- seeking any which can be domesticated and cultured as human food, animal fodder or chemical feedstock.

Returning now to our shrimp culture effort, just to do its front end research required pumping large volumes of seawater through our buildings at up to a thousand gallons a minute. (The commercial farms will have to pump much more.) From the shrimp raceways the water is ejected into wastewater lakes where culture marine algae are grown, removing the excess nutrients before it percolates back into the sea. We can and do cycle some of the marine algae back into the formulated shrimp feeds we have developed.

It occurred to us that having all this pumped seawater on hand was a marvelous opportunity for our plant scientists to initiate some new work in seawater agriculture. Some University of Arizona scientists had tried saltwater irrigation ten years earlier, but only with conventional food crops and only with the usual unsatisfactory results. So we were still not too optimistic about the undertaking.

Seawater Irrigation

In this new endeavor we wished to try something different. If conventional crops cannot be grown in seawater, a more pragmatic approach would be to see what kind of plants can be grown with saltwater irrigation and then determine their utility. We could easily observe that the weedy plants found naturally on desert seacoasts can handle high salinities, high temperatures and high levels of solar radiation. The question was whether any of them would be good for human uses.

We did have some clues. We found it was already known that in earlier times the Seri and Cocopa Indians who roamed the bleak coasts of the Sonoran Desert used almost any seedplant they could get their hands on -- about 80 different species, surprisingly, including some estuarine plants -- for one purpose or another. They couldn't afford to let anything that grew go to waste in that hostile environment. At the same time, and independent of our reflection on this, we were approached by an ethnobotanist from the Arizona-Sonora Desert Museum. He said that he had been told by back country missionaries in Mexico that the Seris' unwritten history has it that they used to harvest a seed crop from a specific coastal halophyte and grind it to make bread. He thought, with our abundance of pumped seawater, we might want to take a look at such a plant. Then, in yet another independent development, we were approached by a scientist from the University of

Delaware, one of the few anywhere who had started culturing halophytes. After visiting our shrimp research station, he enthusiastically urged us to start investigating these strange plants, and offered to join in the effort. This considerable amount of coincidence helped us set our research plans.

Halophytes are unusual plants and there is little known about the details of how they function. They can handle much more salt than any of their related species, and the desert halophytes have to tolerate a burning sun and high air temperatures as well. It is these desert species which have obviously evolved some unusual photosynthetic pathways to avoid destruction from salt and sun. How do they do this?

How Plants Handle Salt

Plants in general can do one of two things when supplied with salty water. Most simply refuse to let salts into their systems at all, which solves the problem rather well. However, this also means such plants are depriving themselves of water, which does little for their survival if there is nothing but salty water available. The second choice is for a plant to take up the water, salts and all, using the water and rejecting the salts, either by extruding them or by isolating them within the plant tissue somewhere they cannot cause harm.

The halophytes we are working with seem to employ the tissue-isolation mechanism. How they manage this is something we would not have been able to understand not many years ago, before it was learned that there is more than one way for a plant to photosynthetically produce carbohydrates.

Most common domesticated food and forage plants utilize the C3 photosynthetic pathway; i.e., they require considerable carbon dioxide, so they must have wide-open leaf pores, even during the hottest period of the day. This in turn permits water to evaporate from the leaves. If there is plenty of water for the roots to take up and deliver to the leaves, this is not too detrimental, since the plants thus achieve evaporative cooling at the same time. To conserve water, however, most halophytes use C4 or CAM pathways which permit them to keep their leaf pores partially or completely closed during the day and open only at night. They thus obtain their required carbon dioxide, but they can reduce their water requirement by more than one-half. For example, a typical C3 plant may need 500 grams of water to produce a gram of dry matter, but a C4 or CAM plant may need only 200 grams for this purpose. This is one of the reasons halophytes manage to survive in salty water; they simply do not need as much water to begin with.

Halophytes also have an ingenious way of handling the salts they do take up. It is believed that the salts are pumped out of the living part of each plant cell and held in the cell's vacuole or waste storage compartment. To do this -- and to thus isolate the harmful salt concentration -- takes high osmotic pressures which mean, in turn, high energy expenditures. One of our plant scientists has calculated that some halophytes spend as much energy coping with high salinities as is required to produce one ton of carbohydrates per acre-year. That's the equivalent of a good yield for many cultivated grain crops.

Their unusual need for osmotic energy may also be a clue as to how some of the desert halophytes put up with high levels of solar radiation. If only processes so profligate with energy can keep them alive in highly saline environments, they need those maximum amounts and intensities of sunlight. Conversely, it suggests that if rugged halophytes from coastal deserts, accustomed to salinities even higher than seawater, had a chance at brackish water much lower in salts, they could use much of that energy for more productive functions, such as faster growth rates and heavier crops.

Questing for Halophytes

There are hundreds of different kinds of halophytes around the world, along seacoasts and in estuarine marshes. They are all inundated with seawater, but they vary widely in response to temperature and wet tropical climates. Many receive some fresh water from rain, dew or ground water intrusions and they have lesser requirements for heat. Few of these latter species will survive under harsh desert conditions. Fortuitously, the research station we use on the Gulf of California offers an ideal environment for growing the desired species. There are less than three inches of rain a year, while the annual rate of evaporation is nearly 100 inches. The only ground water is, if anything, saltier than the ocean, and summer temperatures can hit 44 degrees C (111 degrees F).

We started our halophyte research with some seeds and plants from the University of Delaware, USDA, private seed companies and a couple of scientists in Israel. Seeking the more rugged desert species, we made 15 exploring and collecting trips along the bleak coastlines of the northern Gulf of California and the Baja peninsula. Our scientists found a few dozen native species able to thrive in this rugged

Table 1. Survival and growth of halophytes in outdoor plots watered with seawater, July 1978 to May 1979.

Plant	Survival	Final Size*	Flower/ Seed
Aeluropis littoralis	-	-	-
Aeluropis macrostackyus	-	-	-
Atriplex barclayana	+	21x50 cm	f/s
Atriplex canescans	+	17x41 cm	-
Atriplex cinerea	+	30x38 cm	f/s
Atriplex glauca	+	19x31 cm	f/s
Atriplex lentiformis	+	32x38 cm	-
Atriplex paula	+	27x18 cm	f/s
Barley (salt tolerant)	-	-	-
Batis maritima	+	runners to 70 cm	f/s
Chenoposium album	-	-	-
Distichlis palmeri	+	runners to 100 cm	-
Distichlis spicata	+	little growth	f(male)
Euchenlaena tomentosa	-	-	-
Eucalyptus spathulata	-	-	-
Nitratia schoberi	+	21x4 cm	f/s
Pucinellia distans	-	-	-
Pucinellis stricta	-	-	-
Pucinellia capillaris	-	-	-
Salicornia europa	+	40x35 cm	f/s
Spartina alterniflora (short form)	+	little growth	-
Spartina alterniflora (tall form)	+	little growth	f
Spartina patens	+	little growth	-
Sporobulua airoides	-	-	-

*height times diameter

environment. We ended up with 75 different kinds of halophytes for our first trials.

Our first discovery was that not one of the wild halophytes would germinate in our hypersaline seawater of 40,000 ppm. They had to be started in a little fresh water. This is clearly a survival mechanism, and many desert plants have evolved a similar trick. It guarantees that a new plant won't get started until after a desert rain, when it will have its best chance to stay alive. We found this no real obstacle. For the moment, we germinate halophytes from seed in our greenhouse, using the compact trays developed for forestry seedlings. Extremely little fresh water is required and it is recycled. Once transplanted to the field, the plants are long-lived and may be cropped repeatedly over a course of years before replanting. More importantly, this present fresh-water germination can be described as a temporary inconvenience. When we more fully understand this defense mechanism we hope to devise a way to get around it; such short-circuiting of an unwanted natural behavior is now routine in the plant sciences. In addition, we are continually screening thousands of seeds for greater salt tolerance and to thus attack the problem from a second direction.

Another encouraging discovery was that once germinated, a third of all the species could be weaned quickly to seawater. They were then transplanted to the open desert, and after being irrigated with nothing but seawater for a year, the results looked quite good. As Table 1 reveals, 14 different kinds of halophytes had survived; eight of them did very well, and six completed their life cycle by presenting our researchers with viable seed. Meanwhile, laboratory analyses indicated that some of these plants were high in protein and ought to be directly digestible by ruminant animals. (As much as one-half of the weight of the harvested plants was seed, with a protein content as high as 13 percent and comparing favorably with wheat and soybean.)

Some Promising Contenders

In 1979 we chose 11 of the more promising species and germinated tens of thousands of plants in our greenhouses. These were transplanted into 40 plots on a one-hectare (2½ acres) experimental farm. Table 2 shows that more than 80 percent of them survived the transition to the open desert and irrigation with only highly saline seawater. And, although these plants were laid out during the hottest month of the year and their initial growth was slow, some astonishingly rapid growth followed. We had planned on making our first harvests from the new plantings in the Spring of 1980.

Table 2. Above-ground yield (grams dry weight/m^2) of halophytes on 0.61m centers irrigated every 12, 24, and 72 hours with 40 ppt seawater. Atriplex spp. were transplanted from greenhouse to 18 desert plots (12.3 x 18.5m) on 23 April 1979. Alternated plants were harvested at t = 355 days. Salicornia europaea was sown on 16 February and harvested at t = 265 days.

Water Frequency (hours)	Above-Ground 12	Yield 24	(g/m^2) 72
Salicornia europaea*	1365	**	**
Batis maritima	1137	**	**
Atriplex linearis	927	1134	245
Atriplex barclayana	901	733	431
Atriplex lentiformis	895	620	230
Atriplex glauca	413	348	130
Atriplex canescans	178	165	111
Atriplex polycarpa	143	61	31
Atriplex rependa	22	15	9

*0.31m centers
**Not determined

In fact, growth became so rapid we had to begin harvesting nine of the 11 species in December 1979.

These initial harvests from the one-hectare farm gave us some impressive numbers. The best producers did as well or better than many conventional food crops which have to be irrigated with fresh water. The actual range of above-ground dry-weight annual yield we encountered was 895 to 1,365 grams per square meter. This is as good or better than the numbers for such fresh-water forage crops as alfalfa (annual dry weight yield of 420 to 940 grams per square meter).

The most promising of the halophytes we are working with include several species of salt bush, which can be valuable forage crops; pickle-weed, which is a gourmet vegetable in England; saltwort, which produces edible roots; and a couple of grain crops such as Palmer's grass. (This latter is the spiky little estuarine plant from the Gulf of California from which the Seris may have harvested seeds to make bread -- one of those stories which helped us decide to begin this project. It will be mid-1980 before we will know what grain production may be expected from this grass.) Table 3 lists these promising halophytes.

However promising the lab tests may be, the real questions are whether or not animals can and will really eat halophytes, and if so, what will happen to them. We can't answer those questions yet. The first series of harvests from the one-hectare farm did give us enough seeds and dried plant material for some modest feeding tests with livestock. These preliminary trials are now underway in Hermosillo, Mexico (cattle, swine, poultry) and Tucson, Arizona (goats and smaller farm and laboratory animals); and shrimp feed tests will soon be initiated at the experiment facility on the Gulf of California. We feel pretty confident about the use of halophyte seed crops as a grain feed, so we are really looking to these first trials to indicate how much whole plant material can be mixed into livestock feed without the animals' picking up too much salt. It is believed probable that a livestock diet using a great deal of this plant material would require some sort of preliminary demineralization. Such processes exist.

In addition to the continued growing, harvesting and livestock-testing of our halophytes, some of the larger specimens of the best species have been selected for conservation through vegetative propagation in our greenhouses. Some of these promising species have turned out to be heteromorphic; i.e. the same plant is both male and female. This is most promising, even for plants, because it can enhance and

Table 3. Halophytes planted in field production trials, April-July 1979.

Species	Description	Economic Potential
Atriplex barclayana	Perennial spreading	Forage, grain
Atriplex canescans	Perennial erect shrub	Forage, grain
Atriplex glauca	Small prostrate annual	Grain
Atriplex lentiformis	Perennial shrub to 10' high	Forage, grain
Atriplex nummularia	Perennial erect shrub	Forage
Atriplex patula	Small erect annual	Grain, forage
Atriplex polycarpa	Perennial erect shrub	Grain, forage
Atriplex rependa	Perennial erect shrub	Forage
Distichlis palmeri	Perennial grass with large seeds	Grain, forage
Salicornia europa	Succulent annual	Vegetable crop
Batis Martima	Perennial spreading succulent	Root crop

accelerate breeding experiments to develop new strains for high seed production.

An Overdose of Salinity?

We originally figured that salt accumulation in seawater irrigated soil could be a problem unless we devised some way to cope with it. It seemed obvious that if fresh water irrigation in the desert can cause a salt build-up in the soil, irrigation by seawater would be likely to make things a lot worse a lot quicker. There are indications in the literature that such might not be the case; however, we had to find out ourselves, so we set up complex and precisely measured soil salinity experiments to run in parallel with the growing trials. Some of the plots were irrigated twice a day with saline effluent from the shrimp production facility. Some plots were irrigated but once a day, and some were irrigated only once every three days. At the same time, we used a variety of techniques, including flood and furrow irrigation. We had some preconceptions about the relative effectiveness of these variables, and more often than not, we were wrong.

Irrigating only once every three days left twice as much salt on the surface as irrigating once or twice a day, and plant growth suffered accordingly. Thus, frequent irrigation was indicated. Second, furrow irrigation did not cause any more salt build-up than flood irrigation. Our head plant scientist said he was "astonished" by this. Third, while the seawater from the shrimp raceways had a modest amount of animal fertilizer added to it, it was not enough, we thought, to optimize plant growth. So we tried adding various amounts of other fertilizers to the seawater, and found this did not significantly increase the cròp yield. Now we know that such wastewater from an aquaculture facility is in itself a good source of fertilizer for these halophytes.

Soil type is an important consideration in growing halophytes. We obtained very high infiltration rates when the growing medium was 95 percent sand and five percent clay (unsuitable for any other type of agriculture, of course). This means a lot of the salt in the water leaches right on through and below the plant root zone. On the other hand -- and typical of the surprises these halophytes have had for us -- some of the species we are experimenting with actually grow better in poorly drained soil. This would appear to be of great promise in many inland areas suffering from the problems of saline water.

As a footnote to the experimental results we have just

described, it should be mentioned that we also attempted to grow some of the widely publicized salt-resistant barley which other investigators have cultured on seawater in California. The results were not promising. The barley did at least germinate on full-strength seawater, which is more than the wild halophytes did, but all of the seedlings died within two weeks. Concurrent greenhouse trials showed a much higher tolerance of salt for seed germination than for seedling growth, with the latter falling off when salinity approached a fourth that of our hypersaline seawater. This does not necessarily contradict the work done in California. Our scientists were not using exactly the same strain of barley, and seawater irrigation in the West Coast experiments was applied only once a week. It is probably critical that the California trials were performed in a much more humid climate in a higher rainfall area, and seawater irrigation was necessary only weekly.

Back to Solar Energy

Today, with some uneasiness, we find ourselves returning to the ERL's initial preoccupation with solar energy. We know that results are not very promising thus far in the practical use of solar energy, and we knew ahead of time this would happen. To some, solar energy seems to have moral attributes; because nothing impressive seems to be occurring, they see sinister forces at work. To many others who feel less keenly about it the whole concept of "harnessing the sun" seems a little far fetched, and insofar as they are concerned it may be just as well that nothing is happening.

The real problem, of course, is that solar energy has always been oversold. People get angry or disappointed when it does not produce what they have been led to expect from it, and they have been led to expect too much, too soon. In truth, as a separate fuel source, solar is like oil or coal or uranium ore; all of these are free resources. The costs are in collecting, storing, transporting and converting them. Much solar gadgetry always has been and likely always will be appallingly expensive. So the sinister force at work is simply economics. Many solar techniques -- hopefully not all -- are just too costly.

Like many other researchers with practical experience in this field, we believe that the lowest-cost use of solar energy, for many applications, is, and may continue to be, green plants. Photosynthetic production may not be a highly effective means of solar conversion -- the efficiency is only about one percent -- but green plants are not expensive, and they don't have to be taught how to do the job.

Green plants are able to do something else as well; they are inherently able to offest, if not actually defeat, some of the higher costs of solar development. Consider the fact that in regions which are not highly industrialized, solar hardware is doubly expensive. There is not only the cost of the hardware itself to be amortized, which is bad enough, but there is also the matter of amortizing the conventional fossil energy required in its manufacture. This can be an unsurmountable obstacle in a region which may be limited in both industrial capital and conventional energy and/or may have higher priorities for both.

However, another resource which such capital/energy deficient areas usually have in abundance is people. If a labor surplus is available, the optimum energy development would be something requiring little industrial capital or industrial hardware, but which focuses upon relatively simple, labor-intensive activities.

Planting, growing and harvesting green plants appears to be an ideal solution to this problem. The plant material itself becomes a convenient means of storing energy for subsequent bioconversion processes. Obviously the total investment required is still great and may take many years to amortize. But the more favorable input ratio of human labor to fossil energy would help achieve short- and long-term social objectives as well as help conserve conventional fuel.

Some Geothermal Possibilities

In our view, geothermal energy resources are also a good possibility for parts of the Southwest in general and for agriculture in particular. Apparently Mexico has already moved ahead of us here -- it has quietly been building a good-sized geothermal electric plant south of Mexicali without too much fanfare.

It seems that in our part of the Southwest, on both sides of the border, the great fault lines of the continent run south into the Gulf of California. Indeed, this fault zone has helped create the Gulf. There is a lot of heat locked beneath this ruptured, once volcanic region, and not just along the Gulf itself. From the area around Yuma in a band across middle and southern Arizona, there are (or once were) numerous hot springs and hot water wells; one can still find the names on the maps, even where water tables have gone down so much the visible springs have long dried up. Mining companies and ranches have found hot water when they haven't been looking for it.

Not much has been done with this on the American side of the border, probably because many of our serious geothermalists usually think in terms of high-grade heat, from which electricity can be generated. That's fine, but even some supplies of low-grade heat, around 100 degrees F, could be a bonanza to desert agriculture and aquaculture. Controlled-environment vegetable or fish or prawn production in a frost area would not require expensive conventional winter-time heating. And such water could also be used for halophyte production, downstream from an animal system, as we are now doing in Mexico. The potential of integrated systems for cash crops and animal fodder is intriguing.

Salinity Inland: Problems and Possibilities

Saline water is not confined to the oceans. Salty (and therefore generally unusable and unused) ground water creates problems on all land masses of the world. An important thing to remember about inland saline water is that there is an abundance of it in the United States and Mexico. Hydrologists define saline water as any with a salt content in excess of 3,000 ppm. which is about a tenth as salty as seawater, but still three times too salty for conventional agriculture. The known shallow (within a few thousand feet of the surface) saline ground water areas of the United States total at least 250,000 square miles or one-twelfth the area of the entire nation. If one adds to this those regions with "slightly saline" water -- 1,000 to 3,000 ppm -- two-thirds of the country is included. And in western states like ours, and in much of northern Mexico, many of the areas are virtually unknown, and what are marked off on the maps are simply good guesses.

It is believed that there are more than 18.5 trillion acre-feet of shallow saline ground water in the United States. The "mountain western" states, comparatively short on all kinds of water resources, may contain between two and three trillion acre-feet of this unused resource. It is scattered over about 40,000 square miles.

Within Arizona, known locations of saline water are widely scattered. The largest single chunk of such land is north of the rim country, and the rest of it is spotted throughout the southern half of the state, but mostly in three counties (Maricopa, Pinal and Yuma). It should be noted that in some of these saline locals, conventional agriculture has existed or has been attempted; hence, many fields are now fallow. Indeed, the crop land acreage last reported to be fallow in the three counties was 18.5 percent of the total. (We presume that salinity of ground water is not the only cause of this land idleness, but it is frequently a key factor.)

There is no need to argue further about what might be done with halophitic crops in these inland saline regions. If we can grow useful plants on the seacoast in water of 40,000 ppm, we can certainly do better where the salinity is only one-tenth as great.

The Colorado River Problem

A well-known international dilemma in inland salinity involves the Colorado River. The Colorado and its tributaries are naturally saline from the beginning, leaching minerals from the high country of the southwest and evaporating in the long desert journey to the Gulf of California. Each of the bordering states fights for its share of the river, and there have been near-rebellions over it. California pumps it to Los Angeles for people to drink. The rest of it is used repeatedly for crop irrigation, picking up even more salt. When what is left of the river nears the Mexican border not far above Yuma, salinity is around 900 ppm. By treaty, Mexico is supposed to get 1.5 million acre-feet/year, which is nearly 500 billion gallons. This is a lot of water.

Serious trouble began about 20 years ago, when just before entering Mexico, the river water was diverted once more for a final use in crop irrigation. The salty water became much saltier; salinity at the border went up to 3,000 ppm. Trying to use this to irrigate their own crops, the Mexicans found they were destroying their fields. The Americans put in a diversion canal to drain the worst of the saline water around the Mexicali region, but it was no panacea and the Mexicans remained quite vexed at their loss of land and food production.

Next, the treaty was amended to guarantee to Mexico that its share of the water would be no saltier than the river is upstream from that last big American irrigation district. This means back to 900 ppm again. It also means we have promised something extremely difficult to deliver. Those who promised it had been persuaded that a big desalinization plant on the river could repurify enough of the river to dilute the rest of it to the desired level.

Large-scale desalting was expensive even before OPEC began to reorder the world economy in the early 1970s. Since then, of course, such processes have come to require even more-awesome inputs of capital and energy. Not even the Arabs have put in many units (total world installed

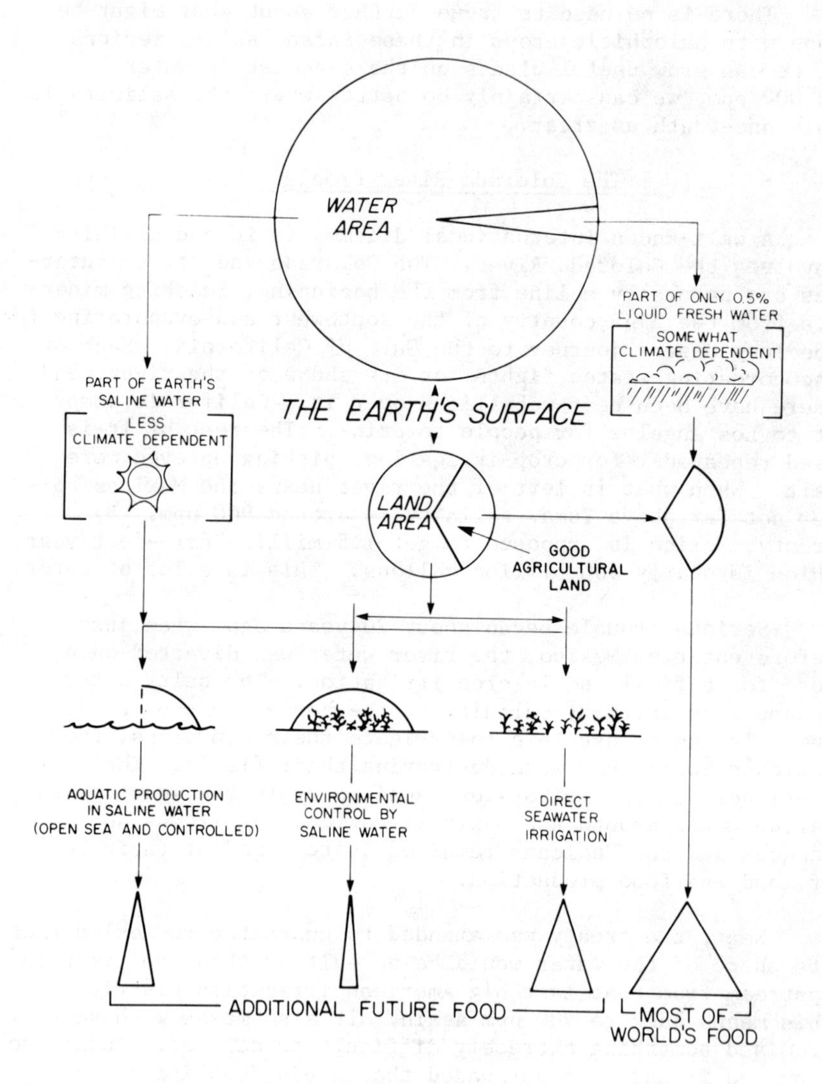

Fig. 4. Conceptual diagram of saline water agriculture in "climate-proofing" food production.

desalination capacity is now about 0.75 million acre-feet), and the Colorado River plant will be the biggest ever built. Some controversy has been inevitable and there have been delays. We understand that present plans call for construction to begin in 1981 and large-scale production to commence in 1984.

About one-third of the projected Colorado River plant capacity will be "blowdown" or wastewater, heavily loaded with the salts removed from the rest of it. The salinity of this wastewater will be about 10,000 ppm or nearly a third as salty as the sea. Something will have to be done with it, and the most obvious something is to canal the blowdown past Mexico's agricultural country to the Gulf of California, just as some of the saltier-used irrigation water is now handled.

As an alternative, we suggest that the naturally salt-tolerant plants we and our Mexican associates are now domesticating as seed crops and forage and energy sources could be irrigated with the saline blowdown water from the desalting station. Halophyte cash crops and livestock feeds could thus contribute to the total effectiveness of the desalting station and perhaps utilize other salty wastewater as well. It is possible that in at least a portion of this region around the Lower Colorado River and its delta, conventional agriculture as it exists could be replaced with new crops. And in such regions, neither we nor the Mexicans would have to worry about the quality of the water at all.

In Summary

The reader may have recognized by now that there appears to be a cohesiveness to all of the aforementioned efforts and thoughts: a curious interdependence of the various separate potentials in saline aquaculture and agriculture, in the use of low quality desert land, in the identification of useful products to be grown in desert intensities of solar radiation, and in all these particular approaches to the challenges of climate-defensive food production. It may well be that a key to the relative economic success of any of them -- i.e. achieving a level of profitability commensurate with desired social objectives, if any -- will be the very integration of several or all in any given application. Fig. 4 represents our attempt at a simple diagram expressing this integrated concept with its potential contribution to "climate-proofed" world food supplies.

Selected Readings

Controlled-Environment Agriculture

Dalrymple, Dana G. Controlled Environment Agriculture: A Global Review of Greenhouse Food Production. U.S.D.A., Washington, D.C., 1973.

Fontes, M.R. Controlled-environment horticulture in the Arabian desert at Abu Dhabi. HortScience 8(1):2, 1973.

Jensen, M.H. and H.M. Eisa. Controlled-environment vegetable production: results of trials at Puerto Penasco, Mexico. Environmental Research Laboratory, University of Arizona, 1972.

Jensen, M.H., ed. Proceedings: An International Symposium on Controlled-Environment Agriculture. Environmental Research Laboratory, University of Arizona, 1977.

Wittwer, S.H. and S. Honma. Greenhouse Tomatoes, Lettuce and Cucumbers. Michigan State University Press, 1979.

Halophytes

Epstein, E. Seawater-based crop production: a feasibility study. Science 197:249-251, 1977.

Glenn, E.P., M.R. Fontes and N.P. Yensen. Seawater irrigation of halophyte crops in the Sonoran deesrt. Submitted to Science.

Goodin, J. New agricultural crops. AAAS Selected Symposium 38, G. Ritchie, ed., 1979.

Hollaender, A., Ed. The Biosaline Concept: An Approach to the Utilization of Unexploited Resources. Plenum, New York, 1979.

Yensen, N.P. and E.P. Glenn. Distribution of salt marsh halophytes on the coasts of the Sonoran desert. Submitted to Journal of Biogeography.

Controlled-Environment Aquaculture

Bardach, J.E., J.H. Ryther and W.O. McLarney. Aquaculture, The Husbandry of Freshwater and Marine Organisms. Wiley-Interscience, John Wiley and Sons, Inc., New York, 1972.

Brown, E.E. World Fish Farming: Cultivation and Economics. AVI Publishing Company, Connecticut, 1977.

Colvin, L.B. and C.W. Brand. The protein requirements of penaeid shrimp at various life-cycle stages in controlled-environment systems. Proc. 8th Annual Meeting, World Mariculture Society, 1977.

Hanson, J.A. and H.L. Goodwin. Shrimp and Prawn Farming in the Western Hemisphere. Dowden, Hutchinson and Ross, Inc., Stroudsburg, Pennsylvania, 1977.

Lightner, D.V. and V.C. Supplee. A possible chemical control method for filamentous gill disease. Proc. 7th Annual Meeting, World Mariculture Society, 1977.

Salser, B.R., L.E. Mahler, D.V. Lightner, J. Ure, D. Danald, C. Brand, N.H.E. Stamp, D.W. Moore, Jr. and L.B. Colvin. Controlled-environment aquaculture of penaeids. Fifth Conference on Food and Drugs from the Sea, 1977.

Stamp, N.H.E. Computer technology and farm management economics in shrimp farming. Proc. 9th Annual Meeting, World Mariculture Society, 1978.

Lloyd E. Slater

12. Synthetic Foods: Eliminating the Climate Factor

Introduction

Despite exciting progress in agricultural science, mankind's food security has become increasingly precarious. Continued and unrelenting variability in climate is at the heart of the problem. Population growth since mid-century has forced most nations to supplement local food supplies with massive imports of grain. When a sequence of bad weather strikes major grain-producing regions of the world, as it did in the early 1970s and to some extent at the end of that decade, world reserves fall dangerously low, food prices rise, and millions more move inevitably towards a condition of hopeless malnutrition.

Modern scientific agriculture, despite its vigorous progress in recent years, is no longer regarded by many as the "open-ended" producer which will inevitably overcome future food supply uncertainties imposed by burgeoning populations and variable climate. Most of the world's good arable land is now under cultivation. Costs for production technology -- for irrigation, new seeds and petroleum-based inputs -- are soaring beyond reach of the world's small farmers.

The combination of variable climate, rising population demand and decreasing resources lends new credence to the notion of furnishing a substantial portion of mankind's food through chemical synthesis in factories, independent of the vagaries of weather and free of the constraints imposed by land and natural environment. What follows is a discussion of the concept of synthetic food production, its status worldwide today, and how this technology will figure importantly in tomorrow's world food system.

Some Definitions

Before proceeding we should define food, nutrient, and

synthesis as used in this discussion, since these terms often have various meanings. We regard food as being any material entering the human diet or feed of animals, even if not metabolized. Nutrients are individual substances that can be metabolized and contribute to growth, maintenance or functioning of the body. Most conventional foods are made up of a number of nutrients (e.g. fats, carbohydrates, proteins, vitamins, minerals). Synthesis is the process of building up a substance from the elements or from simpler substances, themselves capable of being made from the elements. A synthetic food or a synthetic nutrient is a non-living product made by chemical reaction or a biochemical process from non-living raw material. Non-living product is specified since all agriculture is concerned with products made by biochemical processes from non-living raw materials. Under this definition the production of lysine in a culture medium of micro-organism would be termed a synthetic process, whereas the growing of micro-organisms for food would not be called synthesis, but might better be termed microbial farming. Also, foods made by processing natural products without changing their chemical composition are not correctly termed synthetic. Thus, the products made from spun, reconstituted soybean protein and flavored to resemble meats are not synthetic, but simulated meats.

Food Synthesis: A Brief Overview

Synthetic production of naturally grown products is by no means a new and untried method of solving problems of shortages. Since the beginning of the century a succession of items, once derived wholly from agriculture, has been produced by synthesis. Some synthetic products have completely supplanted their natural counterparts, while others are competitive in price or performance.

A significant beginning has already been made in the synthetic production of food by the present day manufacture of items of high unit value that are required in relatively small quantity. Most vitamins added to food products today are made by chemical synthesis, as are an increasing array of flavors and amino acids. These commercial products are exactly the same in composition and properties as products found in nature. In addition, a sizeable industry furnishes many non-natural functional additives to food, such as calcium propionate, which retards spoilage in bread. New developments in synthetic high-energy foods also offer great promise.

These micronutrients, commercially made in factories, already serve significantly in the world food system. The next major need is for the macronutrients -- carbohydrate, fat, and protein. A beginning has been made in the production of each of the basic three. Over the years a number of methods have been devised for making glucose and other edible carbohydrates from cellulose, but none have achieved commercial success. It is to be hoped that the pilot plant demonstration of the production of glucose from waste paper by the U.S. Army Food Sciences Laboratory at Natick, Massachusetts will lead to large-scale production (1). About 100 million kilograms of fats for human consumption were made from coal as the initial raw material in Germany during World War II (2). Fatty acids are now made in large quantities in Eastern Europe by the oxidation of petroleum. They are used for industrial purposes to spare natural fats and oils for food.

The synthesis of food proteins would seem to present great difficulty because the molecules are very large and are made up of combinations of 20 or more different amino acids. Indeed, if one started with amino acids (many of which are now commercially synthesized), the laboratory synthesis required to assemble a protein would involve many thousand steps and unavoidable losses due to low yields. But it may not be necessary to reproduce the molecular structure of the natural substances. The body does not utilize food proteins as such, but only after they have been broken down to the constituent amino acids. In fact, mixtures of amino acids have been used directly as food in the place of protein. They are nutritionally adequate, but unacceptable in the diet because certain of the amino acids have an unpleasant taste. Furthermore, as crystalline, water-soluble substances they contribute no pleasure to eating.

To supply amino acids to the body it would be necessary only to make polymers containing the desired proportions of amino acids that could be broken down during digestion in the same way as natural proteins. Using techniques from modern polymer science, these polymers could be made with desired texture and consistency. While the polymers themselves would be tasteless and colorless, many of the colors and flavors of natural food proteins have been identified and synthesized and several are in commercial production. There is an immediate and urgent opportunity to produce synthetic polymers that simulate and are nutritionally equivalent to the proteins of milk, eggs, meat and fish.

Before concluding this brief overview we should mention that minerals, like vitamins, are of paramount importance because of the great effect of extremely small amounts on nutrition of the body. However, mineral micronutrients present

Table 1. Vitamins in the United States, 1978.

A) Production in the United States, 1978[1]

Vitamins	Production (1000 lb)	Sales (1000 lb)	Unit Value ($ per pound)
Total	37,800	23,749	$ 7.92
Vitamin D	32	12	280.17
Vitamin E	5,839	3,461	16.31
All other	31,929	20,276	23.38

B) Imports of Vitamins into the United States, 1978[2]

Producing Countries	Customs	Dollar Values F.A.S.	CIF
United Kingdom	1,292,222	1,298,230	1,349,345
Others	33,636	35,125	35,125
Total	1,325,758	1,331,876	1,384,470

Table 1, continued

C) United States Exports, 1978[2]

Vitamins	Quantity	Value	Unit Value
Vitamin A & Provitamin A	2,061,864 M UN	$ 1,790,470	$ 0.87
Niacin & Niacinamide	527,461 lb	1,553,015	2.94
Vitamin B_1 bulk	167,703 lb	1,786,319	10.65
D & DL Pantothenic Acid	27,581 lb	127,584	4.63
Vitamin B_{12}, bulk	112,361 lb	2,266,314	20.17
Vitamin C, bulk	1,093,858 lb	3,560,371	3.26
Vitamin E, bulk	586,753 lb	7,474,944	12.72
Vitamins NSPF	1,142,012 lb	7,426,148	6.50

D) Totals--Vitamins in the United States, 1978

	Quantity	Value
Production	37,800,000 lb	
Sales	23,749,000 lb	$ 188,005,000
Exports	5,719,593 lb	25,985,164
Imports	n.a.	1,325,758

1. Source: U.S. International Trade Commission. (Data are available for individual vitamins only when reported by three or more manufacturers.)

2. Source: U.S. Department of Commerce.

Table 2. History of Synthetic Vitamins.

Vitamin		Structure Elucidated	Synthesized	Year
C	Ascorbic acid	1933	Haworth/Karrer	1933
B_2	Riboflavin	1935	Kuhn/Karrer	1935
B_1	Thiamine	1936	R.R. Williams	1936
E	Tocopherol	1938	E. Fernholz	1938
B_6	Pyridoxine	1938	Several groups	1939
	Pantothenic acid	1940	H.K. Mitchell	1940
K	Phylloquinone and menaquinone	1941	E.A. Doisy	1941
	Biotin	1942	V. DuVigneaud	1943
A	Retinol	1931	P. Karrer	1947
	Folic acid	1948	J. Mowat	1948
D	Cholecalciferol	1936		1959
B_{12}	Cobalamin	1950-55	Several groups	1971

Table adapted from Seni (3).

no problem from the standpoint of availability or price. They require no synthetic production, and will not be considered further in this discussion.

Synthetic Vitamin Production

Now a significant world industry, synthetic vitamin manufacture has been hailed as a major advance for better nutrition as well as a signal achievement in chemistry. In the U.S. alone an estimated 19 thousand metric tons of vitamins are made each year. Table 1 shows U.S. vitamin commerce in 1978.

Knowledge on the importance of vitamins in nutrition is still developing. No other substance in the diet, with the possible exception of certain minerals, has so great an effect on the health and well-being of the individual. Small vitamin deficiencies may lead only to impaired efficiency and lowered resistance to infection, but major deficiencies can cause stunted growth, disease, blindness and premature death. The availability of most vitamins in quantity at low cost has virtually eliminated the results of such deficiencies when applied in programs of diet supplement and food enrichment.

Unfortunately, despite low cost availability, vast numbers still suffer vitamin deficiencies. While it is true that a diversified diet of natural foods will usually furnish all the vitamins necessary for normal growth and maintenance of the body, a great many people in all parts of the world do not have such a diet, and hence, would benefit greatly from supplementary vitamins. In some regions the staple foods used by large segments of the population are simply deficient in some of the vitamins. In the affluent Western countries, where vitamin-rich foods are abundantly available, many people still suffer from vitamin deficiencies because of poor eating habits, dislike of certain foods, or gross misinformation spread by food faddists.

Fifty Years of Progress

Although the recognition of vitamins goes back to around 1700 when it was known that fresh citrus fruits would prevent scurvy, the word vitamin did not arrive until 1912, when it was used by C. Funk when he isolated the anti-beriberi factor (Vitamin B-1, thiamine) from rice. But it took until the 1930s for the chemical structures of most of the vitamins to be elucidated, with synthesis in the laboratory shortly thereafter (see Table 2).

All of the vitamins have been made by total chemical

Table 3. Availability and Price of Synthetic Vitamins, 1980.

Vitamin	Chemical Composition	Grade	Units/gram	Lot Size	Price
Vitamin A	Retinol	Pharm.	500,000 A	50 kilo	$27./kilo
		Feed	650,000 A	kilo	$15.40/kilo
Vitamin B_1	Thiamin hydrochloride	USP		100 kilo	$32./kilo
	Thiamin mononitrate	USP		100 kilo	$30./kilo
Vitamin B_2	Riboflavin	USP		25 kilo	$53.-$54./kilo
		Feed		25 kilo	$33./kilo
Vitamin B_6	Pyridoxine hydrochloride	USP		100 kilo	$42./kilo
Vitamin B_{12}	Cyanocobalamin (cryst)	USP		50 gram	$8.-$9.75/gram
	1% w/dicalcium phosphate	USP		25 kilo	$10.75-12.75/kilo
Vitamin C	Ascorbic Acid	USP		100 kilo	$9.90-$10./kilo
Vitamin D	Ergocalciferol and cholecalciferol	USP	850,000 D	kilo	$42.50/kilo
Vitamin E	dl-a-Tocopherol			50 kilo	$29.-$30./kilo
	dl-a-Tocopherolacetate			50 kilo	$24.50/kilo
Vitamin H	Biotin (cryst)			500 gram	$8.50/gram
Niacin	Nicotinic Acid	NF		1000 kilo	$5.25/kilo
	Niacinamide	USP		50 kilo	$6.60/kilo
	Nicotinic Acid	Feed		250 kilo	$4./kilo

Note: Vitamin K, Folacin and Pantothenic Acid are not listed.

Source: Chemical Marketing, February 25, 1980.

synthesis, including vitamin B_{12}, cyanocobalamin, which has a very large and complex molecular structure as may be judged by its empirical formula, $C_{63}H_{88}O_{14}P$ Co. The usual raw materials are either petrochemicals or intermediates that themselves may be made from petrochemicals. Cyanocobalamin, however, is now being produced through fermentation using special strains of microorganisms. One of the simpler vitamins, niacin or pyridine-beta-carboxylic acid, was known as a chemical in 1897, but not identified as a pellegra-preventing factor until 1937.

Synthetic production of vitamins has been commercially successful because relatively small quantities command a high unit price. The cost of extracting vitamins from natural sources, however, would be enormous. For example, Senti (3) has estimated that one would need 6.7 million tons of raw carrots to yield 650 tons of vitamin A (the 1970 U.S. synthetic production). With carrots at 20 cents a pound, the raw material cost alone would be $2.7 billion, not even figuring in labor, overhead, energy and the incredible logistical problem of bringing 335,000 loaded trailer trucks to the plant each year. Considering only the cost of the carrots, the naturally extracted vitamin A would have to sell at $3000 for 1-billion units as compared with today's sales price of around $36 for 1-billion units of synthesized vitamin A.

At the present time pure vitamins range in price from about $10 per kilogram for ascorbic acid to $8 per gram for vitamin B_{12} or cyanocobalamin. Even at these prices the recommended daily allowance of cynocobalamin costs only about one one-hundreth as much as that of ascorbic acid. Over the twenty-year period from 1945 to 1965 the price of ascorbic acid decreased from about $11 per pound to $2, as production increased proportionately. The increase in price since 1965 reflects the current inflation. Table 3 lists the availability and prices for most of the synthetic vitamins in February, 1980.

How They are Used

A large share of vitamin production is taken up today by individuals who customarily take multiple vitamin tablets as a diet supplement. Pauling's popular book (4), which came out in 1970, advocating massive doses of ascorbic acid to prevent and cure the common cold, led to a large increase in the consumption of that vitamin in spite of the contrary medical opinions of some authorities. It is generally agreed that vitamins in moderate excess of the body's requirements are harmless except for the vitamins D.

Table 4. A Case Study in Food Enrichment. A calculation by Dr. A.T. McPherson of the amount of vitamins that would be required to enrich the 1970 rice crop in India.

Vitamin	Amount/Pound of Rice (mg.)	Amount/Crop of 62.5 x 10^9 kg. (1000 kg.)	Value ($1,000.)
Thiamine	2. - 4.	275. - 550.	5,600 - 11,275
Riboflavin	1.2 - 2.4	165. - 330.	3,200 - 6,400
Niacin	16. - 32.	2,200. - 4,400.	5,700 - 11,400
Vitamins F	0.006 - 0.025	0.86 - 3.34	210 - 840

*Data from N. Scrimshaw and A.M. Altschul, 1971. Amino acid fortification of protein foods. M.I.T. Press, Cambridge, Massachusetts.

** Crop estimates from FAO Production Yearbook, Rome, 1971.

In addition to the direct voluntary consumption of vitamins, very significant quantities are incorporated in the diet through the enrichment of processed foods, either in accordance with legal requirements or for purposes of sales promotion. It has long been recognized that vitamins important to nutrition are lost in the polishing of rice and the milling of wheat. Efforts to induce people to eat brown rice or whole wheat bread have been unavailing, so laws have been enacted in many jurisdictions requiring the enrichment of polished rice, white flour, and other cereal products to replace the vitamins lost in processing.

To furnish some idea of the amounts of synthetic vitamins required in a practical situation, Table 4 was prepared by Archibald T. McPherson* to show the quantities and value of vitamins needed to enrich the entire 1970 rice crop of India in accordance with maximum and minimum standards set up by the U.S. Food and Drug Administration. The crop that year, estimated at 62.5×10^9kg was about one-fifth of total world production.

Thus, the value of the vitamins required for maximum enrichment of the entire 1970 rice crop would have been about $30 million at that time, representing about one-third of one percent of the value of the crop, taking the price of rice as $0.15 per kg. Producing this amount of vitamins would not tax the available manufacturing facilities, since the amount of vitamins required was only about 65 percent of U.S. production that year, with the U.S. being only one of a number of producing countries.

Well-Nurtured Livestock

During the past few decades animal nutrition has become an increasingly exact science, with large quantities of vitamins and other nutritional supplements now being used in feeding livestock and poultry. Grains, oilseeds, forage, and other feedstuffs are analyzed, and vitamins, amino acids, and minerals added to correct any deficiencies. The composition of the ratios is determined by computer to a fraction of a percent according to the age, sex and purpose of the animal, so as to yield maximum return in milk, eggs, or meat.

With a few exceptions, animals require the same vitamins as man. In fact, the young of all species require vitamin A in amounts that are roughly the same at comparable stages of growth. The recommended daily allowance for a child at 20 percent of his adult weight is 150 International Units of

---

*See Acknowledgement at end of chapter.

vitamin A per day per kg of weight, while that for a chicken, also at 20 percent of mature weight, is 200 units per kg. All species require ascorbic acid, but only man and the other primates and possibly the guinea pig must obtain it in their food, the other animals being able to synthesize it in their bodies. Choline and pantothenic acid are almost never deficient in the human diet, but these vitamins must be added to the feed of poultry and swine for optimum growth when they are raised in confined space.

Synthetic Fats and Oils

Fats and oils play an important role in human nutrition. They have a high energy content, supplying about 9.3 kal/g, which is over twice that furnished by carbohydrates and proteins. However, the calories supplied are less readily available since fats and oils are ingested and stored in the body rather than quickly metabolized. Fats and oils also serve as carriers for the fat-soluble vitamins A and D and are a chief source of vitamin E. They are the source of the essential fatty acids required for structural development of tissues and prevention of fat deficiency disease such as eczema.

At present fat and oil consumption is clearly linked to affluence. People in the poorer, underdeveloped countries gain only about 14 percent of their calories from fats and oils, while those in Western countries average more than 30 percent, with more than 40 percent for the United States. In any case, people everywhere have a strong preference for fats and oils in their diet, particularly when associated with meat and dairy products.

In view of the strong desire for more fat and oil and the high cost for producing the natural product when compared to less expensive sources for calories, it seems likely that synthetic fat, if it can be made cheaply enough, could be an important future source of the world's calorie budget.

Fats were made by synthesis for human consumption on a large scale in Germany under the emergency conditions of World War II. The initial raw materials were coke, air and water, with the method employed being the well-known Fischer-Tropsch reaction, as shown in Figure 13 in Chapter 7. Coke from coal, heated in the presence of steam gave a mixture of carbon monoxide and hydrogen -- the familiar water gas reaction. In the presence of catalysts, the water gas, with or without the addition of hydrogen, gave a mixture of hydrocarbons having a wide range of molecular weights. Liquid hydrocarbons thus obtained were used to replace the gasoline from petroleum. A waxy fraction known as "gatsch" was converted to fatty acids by catalytic oxidation with air. After purification these fatty acids were combined with glycerine to form fats, the glycerine having been made synthetically from low molecular weight hydrocarbons. The fats were then freed from undesirable smell or taste and converted to margarine. About 100 million kilograms of fat were manufactured in this manner. Production was not continued after the war because the price was not competitive with that of natural fats and oils.

Schubert (2), in reporting on the quality of the wartime synthetic fat, noted that "the Germans, not having anything else, thought the product was satisfactory...(but) the British who examined it at the end of the war thought it was terrible." He went on to note that part of the problem was that the product really did not match natural fats. In butter, for example, the number of carbons runs from 4 to 16, with all even-numbered, whereas the manufactured product had an equal percentage of odd- and even-numbered fats. Also, about one-quarter of the synthetic product was branch-chained, while natural fats are unbranched. However, Schubert felt that chemical engineering, if the will is there, would be able to deal with this problem today.

Improvements Ahead

Frankenfeld and associates (3) exhaustively reviewed the German process and conducted a critical study of possible means of synthesizing edible fatty acids from respired carbon dioxide on board a space craft. They concluded that the only promising route involved converting carbon dioxide to carbon monoxide, reducing the carbon monoxide to ethylene, polymerizing the ethylene to alfa-olefins via the Ziegler growth reaction, and then converting the high molecular weight olefins to fatty acids by oxidative ozonolysis, followed by combination with glycerol to form fats. An engineering study of the system showed it would be very complex and difficult to control, so its further development was not recommended. In the course of the work, however, methods were found for the synthesis of glycerol that might be practical for use on a spacecraft, so the use of glycerol as a source of energy as an alternative to fat was recommended.

The production of synthetic fat from petroleum appears simple. It is only necessary to oxidize a suitable fraction of petroleum by air at a somewhat elevated temperature, using a catalyst such as potassium permanganate, and then to separate and purify the resulting fatty acids and combine them with glycerol. In practice, difficulties arise in various side reactions that make it impractical to carry the oxidation of a given charge of petroleum to completion in a single stage. Furthermore, problems are encountered in separating and purifying the fatty acids. As Schubert points out above, however, major concern arises from the fact that in the product, like the petroleum fraction from which it is made, is a mixture of fatty acids of even- and odd-numbered carbon chains, whereas natural fats contain only an even number of carbon atoms in the fatty acid molecule.

So far as can be determined, there is no significant production of synthetic fats and oils for food in the world today. Synthetic fatty acids are, however, made in the Soviet Union in large quantities for industrial purposes, in order to free natural fats and oils for human consumption. In 1969, for example, the Soviet output of fatty acids amounted to about 270,000 metric tons. At that time Russian consumption of oils, fats and greases was about 33 grams per person per day. Thus, the quantity of fatty acids made industrially spared about 294,000 metric tons of fat, or the amount consumed by about 240 million Russian people.

The supply of natural fats in the United States is adequate for producing fatty acids and their derivatives for industrial purposes. Nevertheless, the American Oil Chemists' Society has been concerned about the possible threat to this production which is posed by synthetic fatty acids, as indicated by a symposium held on the subject in 1966 (6). A review of new developments in synthetic fatty acids by Sonntag (7) considered the relative advantages of making lauric acid from coconut oil and synthesizing it, and indicated some factors favored synthetic production.

Fats are scarce and relatively expensive in some of the developing countries. As a consequence, in some places special measures are required to keep motor oil from being used as an adulterant or otherwise mistakenly consumed for food. Thus, fats should be marked for early consideration in any program for the synthetic production of food.

Synthetic Carbohydrates and Other Energy Foods

Carbohydrates are the most substantial part of the human diet, furnishing energy at the rate of 4 kcal per gram, as well as structural elements within the body. Composed of carbon, hydrogen and oxygen, with two atoms of hydrogen for every oxygen, they include polysaccharides such as sucrose, starch and glycogen, which are digested to glucose, and simple sugars such as lactose and fructose. They also include substances such as cellulose, pectins, lignins and gums which are undigestible by man, but can be metabolized by ruminant animals. They are synthesized in nature by plants, combining carbon dioxide and water in the presence of green chlorophyll through the energy furnished by sunlight.

Plant-produced carbohydrates are the world's most abundant food constituent, so one would expect they would be the last to be considered for practical synthetic production. Nevertheless, as a spin-off from their studies of closed life-support systems in space craft, the astonishing projection was made by Weiss of Worcester Polytechnic Institute and Shapira of the National Aeronautics and Space Administration that the manufacture of edible sugar could be economic and capable of contributing to the world's food supply by 1985 (8). They went on to say that eventually it may be possible, using inexpensive formaldehyde as the substrate, to produce sugars as cheaply as from $0.07 to 0.11 per kilogram.

Weiss detailed the synthetic process for edible sugars in a paper he presented in 1976 at the First International Congress on Engineering and Food (9). The scheme for producing metabolizable sugars from formaldehyde is shown in Figure 1. It involves the calcium-hydroxide-catalyzed polymerization of formaldehyde to formose sugars with their subsequent separation and purification by chromatographic means. Weiss offered several provisos on this method. One was that he found it "difficult to project that chromatographic separation of carbohydrates will eventually be an economic process." He also pointed out that the utility of formose sugars depends on two factors: a) the reduction or elimination of toxic branched species so that the sugar is not toxic to humans or lower organisms, and b) the cheap production of formaldehyde so that it is economically advantageous to use formose instead of natural sugar.

During the same 1976 Congress, which included a special session organized by the author on "The Present and Future of Synthetic Foods," Spano of the U.S. Army Natick Research and Development Command, Food Sciences Laboratory, reported promising work in making carbohydrates through the enzymatic hydrolysis of cellulosic waste material (10). Figure 2 shows the pilot process that has been under development and test at the Natick facility for several years. The first step is production of the enzyme from the fungus *Trichoderma viride*. The enzyme broth, filtered from the culture-growing medium, is combined with milled cellulose waste, with saccharification taking place at atmospheric pressure at a temperature of 50°C. The unreacted cellulose and enzyme are recycled back into the vessel and the crude glucose syrup is filtered for

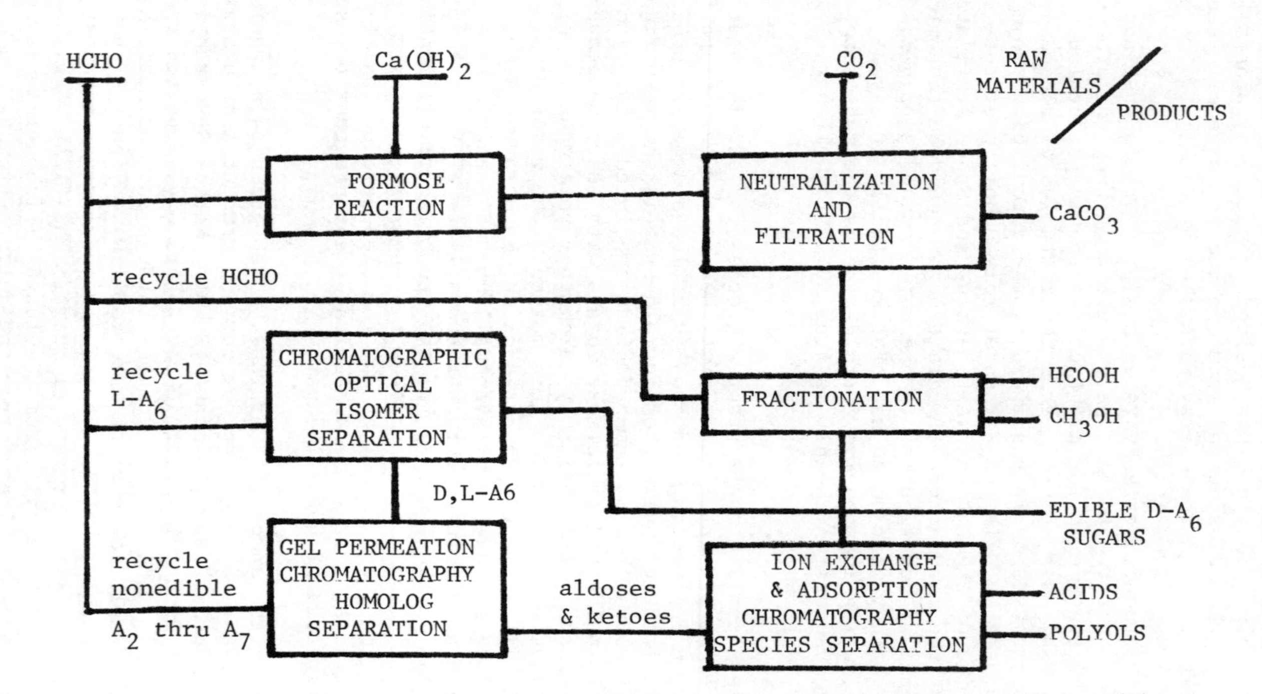

Fig. 1. Flow sheet for the production of sugar from formaldehyde. Weiss (9).

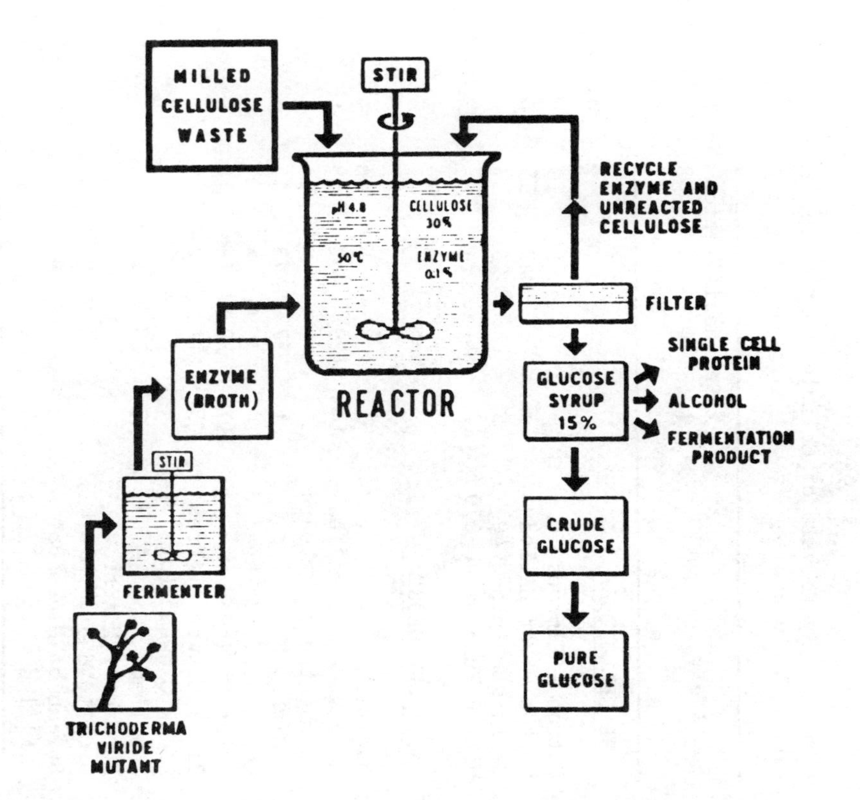

Fig. 2. Pilot process for the synthesis of glucose from cellulose waste by enzymatic conversion at the U.S. Army Natick Laboratories. Spano (10).

Table 5. Hydrolysis of Cellulose by Trichoderma Viride Cellulase.

Substrate	% Saccharification			
	1	4	24	48
	(hours)			
PURE CELLULOSE				
Cotton - Fibrous	1	2	6	10
Cotton - Pot Milled	14	26	49	55
Cellulose Pulp SW40	5	13	26	37
Milled Pulp Sweco 270	23	44	74	92
WASTE CELLULOSE				
Bagasse	1	3	6	6
Bagasse - Pot Milled	14	29	42	48
Corrugated Fibreboard Mighty Mac	11	27	43	55
Corrugated Fibreboard Pot Milled	17	38	66	78
Black Clawson Fibres	5	11	32	36
Black Clawson Pot Milled	13	28	53	56
Bureau of Mines Cellulose	7	16	25	30
Bureau of Mines Pot Milled	13	31	43	57

use in chemical or microbial fermentation processes to produce single cell protein, alcohol for fuel, solvents, etc.

While the Natick pilot plant has mainly used waste newspaper as its cellulose source, many other industrial and agricultural wastes have been evaluated as substrates for similar hydrolysis, as shown in Table 5. Final percent saccharification in the reactor ranged from 33 to 77, increasing with enzyme activity, but decreasing as the substrate concentration increased. The estimated cost for producing glucose was 11.3 cents per pound. A plant producing 100 tons per day would supply the carbohydrate requirements for 250,000 people.

Non-Natural High-Energy Foods

In a search for synthetic substances that might provide a higher caloric density than natural carbohydrates, Miller (11) selected two as models for study; one was the commercially available 1,3-butanediol, and the other was 2,4-dimethyl heptanoic acid which had been synthesized for the investigation. Feeding studies showed that these substances were metabolized by animals and supported the conclusion that it is possible to design and synthesize substances capable of being used by the animal for energy.

Subsequent investigations were carried out by the U.S. Army Natick Laboratories to compare butanediol, glycerol and propylene glycol with fat as sources of energy. Observations were made of the performance of animals under conditions of stress from exposure to cold, and from exercise to exhaustion. The authors found that the synthetic food compounds were utilized by the animals as a replacement for fat under a variety of conditions and concluded that such compounds may be important components in military and civilian foods of the future (12).

Synthetic Proteins

Proteins are crucial in all living organisms and play many roles in structure and function. They are the primary organic base in skin, muscle, cartilage and hair. All of the enzymes and most hormones are proteins. Chromosomes are highly complex proteins, and the mechanisms responsible for oxygen and electron transport (hemoglobin and the cytochromes) are conjugated proteins. Meat, fish, eggs and cheese are foods often referred to as "proteins" because of their high amino acid content.

Proteins are large polymeric molecules which invariably contain nitrogen, carbon, hydrogen and oxygen; almost invariably sulfur; and often, phosphorous. All the proteins in nature are initially produced by photosynthesis in the green plant -- the green leaf on land and algae or phytoplankton in water. Plant proteins may be stored in concentrated form in seed and in a much less concentrated form in tubers and roots.

The animal does not manufacture protein but rather serves as a converter of the protein in plant products used as feed. It concentrates the plant protein and changes it into the variety of animal products that are used for human food. The larger part of the protein consumed by the animal is used for its own metabolic purposes. The fraction of the protein in feed that is converted to products employed as human food may range from 4.5 percent for beef to as high as 30 percent for milk. Efficiencies of 20 to 23 percent are reported for eggs, 18 to 26 percent for poultry meat, and 12.5 to 19 percent for pork.

Structurally the protein molecule is a long chain made up of primary amino acid groups. The amino acid, as the name implies, combines both acidic and basic functions in the same molecule. Thus, amino acids readily form long chains by the combination of the amino group of one amino acid with the carboxyl group of another. There are about 20 different amino acids in natural proteins. Joined together in different numbers and in different sequences, they give rise to a great variety of proteins. Proteins differ from polypeptides in that molecular chains containing more than 100 amino acid groups are arbitrarily called proteins, and those containing fewer, polypeptides.

In the process of digestion proteins are broken down into the constituent amino acids; hence, it is the number and kind of amino acids that are significant, not the protein molecules as such. Eight of the amino acids (nine for children) are termed essential because they must be supplied individually to the body, whereas the other amino acids can be converted by the body from one to the other and used interchangeably. A very significant feature of the essential amino acids is that they can be utilized for growth or maintenance only to the extent that they are present in definite proportions, each to the others. If only one should be present in less than the requisite proportion, the other essential amino acids can be employed in growth and maintenance only to the extent that the limiting amino acid is present. Lysine, threonine, and tryptophan are limiting in most cereal grains and methionine in oilseeds. Lysine, present in two-thirds of the optimum proportion, is the first limiting amino acid in wheat. This means that only about two-thirds of the other essential amino acids can be used to build or repair tissues; they can, however, be metabolized for energy.

Measuring Protein Quality

Owing to wide differences in the composition and availability of proteins, the accurate measurement of their nutritional value is important at all stages -- from the selection of a diet by an individual to the planning of the world food supply by an international organization such as FAO. The many factors involved in protein metabolism make the measurement of quality difficult and complex. Chemical, microbiological, biological, and clinical methods have all been used. In a comparison of methods Swaminathan (13) found that the Protein Efficiency Ratio may serve as a simple, convenient index, and that it is in reasonable agreement with the more sophisticated measurements. This ratio, abbreviated as PER, is simply the gain in weight in grams of a young animal per gram of protein in the food that it consumes. The value found for the PER of a given protein may vary with age and strain of the rats commonly used for the test, as well as with the environment and the level of protein in the feed. Hence, the results recorded in the literature may show a considerable spread for the same protein. To make the results of different studies more closely comparable, some investigators now conduct parallel tests with standard casein. By assigning a PER of 2.50 to the casein, all results can then be corrected to a common base.

Average values of the protein efficiency ratios of some common proteins are shown in Table 6. When the proteins from animal and plant sources are arranged in descending order of PER values, the animal proteins all lie in a range above that of the plant proteins and there is no overlap. Whole egg has the highest PER with a value of 3.92. Soybean, with a PER of 2.32, is the highest of the plant proteins.

The high values of PER in animal proteins indicate that they contain the essential amino acids in very nearly the proportion required by the body. The cereal grains are quite deficient in lysine, and to a lesser extent in tryptophan, except for the new high-lysine maize. Oilseeds and pulses, on the other hand, contain an adequate proportion of lysine but are deficient in methionine, which is present in reasonable amount in the cereal grains. Thus, their amino acid patterns are complementary so that mixtures, as shown in the table, exhibit higher values of the PER than either constituent alone. There is nothing new, of course, about the use of combinations of plant foods to furnish protein that is reasonably adequate for the human diet. Indeed, such diets are

Table 6. Protein Efficiency Ratio (PER) of Some Animal and Plant Proteins.

Source of Protein		Average PER
Animal:	Egg (hen's, whole)	3.92
	Fish	3.55
	Milk (cow's and buffalo's)	3.09
	Casein, purified	2.86
Plant:	Soybeans	2.32
	Oats, as cooked oatmeal	2.25
	Chick peas	1.68
	Groundnuts (peanuts)	1.65
Mixtures:		
	Soybean flour PER=2.32, 36% + Rice PER=1.41, 54%	2.88
	White beans PER=1.40, 20% + Maize PER=1.41, 80%	2.04
Cereal Grain fortified with amino acids:		
	Whole wheat	0.93
	Whole wheat + 0.2% L-lysine HCl	1.45
	Whole wheat + 0.2% L-lysine HCL + 0.2% DL-threonine	2.00
	Whole wheat + 0.3% L-lysine HCl + 0.3% DL=threonine	2.44

Data from FAO Nutritional Studies No. 24, 1970. Amino acid content of foods and biological data on proteins, FAO, Rome.

traditional in many different cultures. However, during the past few decades the increasing use of individual cereal surplus grain, mainly wheat, in national emergency food programs has threatened and often eliminated such traditional diets.

Modern nutritional science can guide the selection of grains, oilseeds and pulses from among locally available products to obtain the highest PER that is economically practical. The PER can often be raised further and brought into the lower part of the animal protein range by the addition of small fractions of a percent of one or more synthetic amino acids. Numerous clinical studies have shown that even severe childhood malnutrition manifested by kwashiorkor or marasmus can be cured by various upgraded plant proteins without the addition of milk, eggs, or meat to the child's diet. This, too, is not altogether new. Countless children have been nourished more or less adequately by diets of cereal grains and pulses even though the practice is not used very effectively in developing countries today. Now that plant proteins can be upgraded to somewhere near the animal protein range, the proposal has been made in many quarters that we discontinue the feeding of cereal grains and oilseeds to livestock and poultry and use them instead for direct human consumption, to buy time in which to bring the population into equilibrium with the food supply.

World Protein Needs: 1985

Before discussing the synthetic production of protein it is important to consider other present and prospective means of production, and to estimate future supply from these sources. By so doing it will be possible to estimate the quantity of synthetic protein that may be needed to fill the gap and supply a truly adequate amount of high-quality protein for all of the world's people.

Let us take the year 1985 as a target date, since that is the year the Food and Agriculture Organization of the United Nations earmarked, in estimates made in its Provisional Indicative World Food Plan for Agricultural Development (14). This plan embraces all food products, but we will consider here only those parts concerned with protein. These are presented, for the most part, in that document's Chapter 5 on "Targets, Policies and Inputs for Livestock Production," and Chapter 13 on "Consumption and Nutrition Perspectives and Food Policies."

Between the base year 1962 and the target year 1985 FAO hopes to see the production of meat and milk doubled and egg production increased almost three times. Fish production is

Table 7. Supply of Animal Protein for Developing Countries, FAO Plan.

Product	Base Year 1962 Population- 1,395 million			Projected Year 1985 Population- 2,515 million		
	Quantity (1000 m.t.)	Per person/day Amount (g)	Per person/day Protein (g)	Quantity (1000 m.t.)	Per person/day Amount (g)	Per person/day Protein (g)
Meat**	15,330	30.15	4.83	31,362	34.15	5.45
Milk**	59,651	117.	4.09	113,850	124.	4.34
Eggs**	1,505	2.96	0.39	4,410	4.8	0.52
Fish, dressed**	4.500	8.9	1.64	13,150	14.5	2.67
			10.95			12.98

Source: Provisional Indicative World Plan for Agricultural Development, FAO, Rome, 1970, Volume 1, p. 23 and p. 271.

**The protein contents of meat, milk, eggs and fish are taken as 16, 3.5, 13 and 18 percent, respectively, and the dressed weight of fish, as one-half of the live weight.

also to be greatly increased and its consumption as human food more than doubled. The quantities of these products scheduled for 1985 are shown in Table 7 together with the estimated protein content of each. The increase in production between 1962 and 1985, if accomplished, would indeed be a very significant achievement. But at the same time the population will have increased in about the same proportion, so that the gain per person will be small, as shown in the table. The total amount of animal protein per person per day will be about 13 grams as compared with 70 grams, or more than five times as much as currently consumed in the United States.

The actual amounts planned by FAO can be better visualized if put in terms of everyday food products, as shown in Table 8. It is, of course, sheer whimsy that each person in the developing countries will get each day a small part of a hamburger, a cup of milk, and one-twelfth of an egg. Most of the people will get no animal protein most of the time, while a few will dine as we do in the United States.

In our new era of rising expectations there is a real question as to whether the underpriviledged two-thirds of the people of the world will continue to accept this wide disparity in nutritional levels by the year 1985. The nutritionists cannot answer this question, but must refer it to the social and political scientists.

The wide disparity in quality and quantity of diet may be reduced somewhat by new sources of high-quality protein now under development. The upgrading of plant protein by synthetic amino acids has already been mentioned. Production of Leaf Protein Concentrate from crops such as alfalfa may render it possible to obtain more protein per acre for human food than from grain, oilseed, or pulses. Single Cell Protein is also under development in nearly all industrialized countries. The output, now on a large experimental scale, is mainly being used in animal feeding. A related development is the texturizing, coloring and flavoring of soybean protein to resemble meat. While products of this kind are gaining wide use and acceptance, it should be noted that they must be considered simulated, not synthetic meat.

World demand for more and higher quality protein is such that all of these products will be used to the fullest extent practical. While no one of them duplicates the nutritional characteristics of animal protein, singly and in combination they can furnish human protein (and amino acid needs). Three are dependent on agriculture, and hence would not extend the supply of protein if population demands should exceed the

Table 8. Every Day Equivalents of Animal Protein Planned by FAO for 1985.

Item	Amount per persons per day (g)	(oz)	Portion of a Common Food Item
Meat	34.15	1.2	Three-tenths of a hamburger
Milk	124.	4.4	One-half cup of milk
Eggs	4.8	0.17	One-twelth of an egg
Fish	14.5	0.5	One small sardine

capacity of agriculture to produce. In the long run, then, the choice lies between Single Cell Protein, a less than optimum product, and proteins fully equivalent to animal protein -- yet to be made by total synthesis as proposed in this discussion.

Why Not Amino Acids?

All 20 amino acids that constitute food proteins have been made by synthesis and are commercially available in lots of at least one kilogram. So before considering the synthesis of protein, the question might be asked, "Why not simply incorporate amino acids directly into the diet?" A practical consideration is that the pure, crystalline amino acids would have no organoleptic or "taste" appeal; in fact, some of them as now manufactured have an unpleasant taste. A further reason is that amino acids may be better assimilated if released slowly by digestion rather than if provided all at one time.

Another major problem with synthetic amino acids lies in the fact that chemical synthesis produces equal amounts of the dextro- and laevo-form, whereas the body utilizes only the laevo. With methionine and phenylalanine the body is able to convert the dextro- to the laevo-form, but with other amino acids the dextro-form can only be used, if at all, as a source of energy. Means are available for separating the isomers and converting the dextro- to the laevo-form, but the difficulty and cost of the additional steps seem to have limited their use. At present it seems more practical to make lysine by fermentation, a process which yields only the laevo-form.

Progress in Protein Synthesis

The first synthesis of a true protein was accomplished early in 1969, although polypeptides, similar in structure but of much smaller molecular size, have been synthesized in considerable number during the past 100 years. The enzyme ribonuclease was synthesized by two groups of investigators at about the same time, though they were working independently: Gutte and Merrifield of Rockefeller University (15), and Denkewalter, Hirschman and colleagues at the Merck, Sharpe and Dohme Research Laboratories (16). Both groups of investigators confirmed the identity of the synthetic ribonuclease by demonstrating its enzymatic activity. The structure of this enzyme had been shown previously to be that of a long chain molecule made up of 124 amino acid groups, the identity and sequence of each having been determined.

At Rockefeller University the procedure was the solid-phase, synthetic peptide route developed by Merrifield (17). In this procedure the terminal amino acid is firmly attached to a microscopic polystyrene bead and each of the other 123 amino acid groups is added, one at a time, by appropriate reactions. Attachment to the polystyrene bead keeps the product insoluble and greatly facilitates removal of successive reagents by washing. After the molecule has been built up it is separated from the polystyrene bead. In making ribonuclease the process was automated and controlled by a computer so that the 369 chemical reactions and the 11,911 individual steps were accomplished in a few weeks.

The Merck Laboratory team used a more conventional procedure. They first synthesized a number of small portions of the molecular chain, which were then joined together to form larger subunits which, in turn, were combined to yield the desired molecule.

In looking at these procedures one might conclude that any practical manufacture of proteins would be impossibly difficult and expensive. Such would be true if the intent were to duplicate the structure of natural food proteins by either of the foregoing methods. It might, however, be possible to make equivalent products by conventional polymerization methods. The essential requirement would be: (a) that the proteins would contain essential amino acids in the requisite optimum proportions, along with suitable amounts of non-specific amino acids, and (b) that these proteins would be capable of digestion to yield the constituent amino acids at about the same rate as for natural proteins.

Evolutionary Protein?

A wholly different approach is suggested by research on the production of proteins in connection with the beginnings of life on earth. It has been shown that polymers containing most, if not all of the amino acids in proteins, can be formed from simple, gaseous substances that may have been present in the atmosphere of the primitive earth. One of several comprehensive sources of information on this subject is the excellent treatise by Calvin on "Chemical Evolution" (18). During the 1976 Congress on Engineering and Food, a paper on the so-called pansynthesis method of producing a mixture of amino acids, entitled "Prebiological Chemistry: A Model for Synthetic Food" was presented by Cyril Ponnamperuma of the University of Maryland. (Unfortunately, this paper is unpublished and unavailable.)

Experiments have been conducted with nitrogen, methane,

water vapor, formaldehyde, hydrogen cyanide, and acrolein in various combinations under exposure to heat, light, electric discharge, and radiation from radioactive sources. The reactions that have been observed take place in a single stage and yield for the most part very small amounts of polymers that are made principally of the simpler, non-specific amino acids. These experiments foreshadow the possibility that conditions and combinations of reactants might be found that would yield a usable protein in a single stage operation. The prospects of success seem small at the present time, but sustained research is clearly warranted in view of the tremendous value such a discovery would have. Fortunately, such research need not be conducted in a trial and error manner. The application of thermodynamics may serve as a guide for the selection of materials and conditions that could lead to a breakthrough.

Prospects for Synthetic Foods

While the increasing impact of variable climate on world food supplies and the nearing limits of conventional agricultural production are compelling reasons, other timely factors support a serious world effort to develop and commercialize synthetic foods.

First and foremost is the rapidly advancing science of nutrition and the creative role that synthetic food components, combined into well-balanced, nutritionally "tailored" whole foods, could play in ameleorating malnutrition worldwide. Much has been learned, for example, about the digestion of proteins and the metabolism of amino acids. It may well be that foods containing synthetic proteins having different patterns of amino acids will be found desirable for optimum nutrition at different periods in life.

Organic chemistry has reached a stage where nearly any desired substance can be synthesized from available raw materials by a practical route. The chemical engineer can then produce the substance at a competitive price if the demand is sufficiently great to warrant the automation of manufacturing facilities. Polymer science is now in a position to provide the means for making synthetic protein molecules in sizes and configurations best suited to different applications, such as the production of fibers, gels, dispersions, or brittle products.

Engineered Foods

Progress in food science and engineering is furnishing the technique for food that can be designed and manufactured from synthetic components. Oleomargarine, the first food product to be created from components to simulate natural butter, has been so successful that butter is now being modified to simulate some of margarine's desirable qualities. Simulated milk, made from plant ingredients, is now competing with cow's milk in nutritional quality and taste appeal, yet is significantly lower in cost. Meat analogs, fabricated from textured vegetable proteins and blends of basic ingredients, are being sold as breakfast foods low in cholesterol and with improved cooking qualities. A good review of the status and possibilities of food engineered from macronutrient components and ingredients is offered by Inglett (19).

Use of synthetic food components should bring new precision, economies and nutritional benefits in the manufacture of engineered foods. Compositional variation in natural ingredients due to varietal differences and growing conditions will be eliminated in the synthetic product. Undesirable constituents in natural foods, such as toxic chemicals, biodegredation enzymes, undesirable flavors, can be "designed out" in the product engineered from synthetic components. The best possible product balance between nutrition, palatability, convenience and cost to the consumer can be achieved, as suggested by Slater in 1978 (20).

Materials for Food Synthesis

Under present conditions the basic raw materials required for large-scale synthesis of food would be one or more of the fossil fuels, water and air. In addition, sources of sulfur, phosphorus, and minerals would be needed to synthesize certain essential nutrients.

Petroleum has been the principal source of carbon in organic synthesis to date, but it could be replaced in food synthesis, at some added difficulty and cost, by coal or even wood. However, even if petroleum should be used in food synthesis, the amount required would be only a very small portion of what is now wastefully used as fuel. In 1975 the world production of liquid and gaseous petroleum had an estimated energy content of 35.1×10^{15} Calories. All the food consumed in the same year had an estimated energy content of 3.5×10^{15} Calories, or only about one-tenth as much. Producing enough synthetic food to close the gap between the 3000+ Calories per person per day in the rich countries and the 2000 to 2400 Calories of the poor countries would require 0.7×10^{15} Calories, or about two percent of world petroleum production.

A more precise idea of the relation of the production of

fossil fuels to the world food supply is shown in Table 9, which compares their respective quantities and energy values. When the fuel is expressed in terms of coal, crude petroleum and natural gas, and the food is given as protein, fat and carbohydrate, the weight of the food is 14 percent of the weight of the fuel, and the energy value of the food eight percent that of the fuel. Assuming that a pound of fuel would produce a pound of food, the synthesis of all the food in the world would not make a very large increase in the consumption of fuel.

It should be noted that even this projected consumption of petroleum and other fossil fuels to produce synthetic food would be offset by their use in modern agriculture as fuel for farm machinery, for the production of fertilizer and pesticides, and for drying of crops and irrigation. In many situations today the energy consumed is greater than the energy content of the food grown.

Nitrogen in the form of ammonia is used both as fertilizer and for the synthesis of amino acids and proteins. Most of the current production of ammonia is from nitrogen of the air, natural gas, and water, though some is made with the use of coal or liquid petroleum instead of natural gas. Looking to possible future dependence on the biomass as a renewable source of both materials and energy, methane produced by fermentation could be used instead of natural gas for production of ammonia.

When ammonia is used for the synthesis of amino acids and polypeptides and protein, there is no significant loss of material, whereas when ammonia is applied to the soil as fertilizer only a small fraction of the nitrogen is returned as food at the end of the food chain. It has been estimated that for the year 1969 in the United States the amount was 15.6 percent (McPherson, 1972). The recovery of other elements was even less -- 8.7 percent for potassium and 5.5 percent for phosphorus. Thus synthetic production has a conspicuous advantage over conventional agriculture in the saving of materials and the energy consumed in their production.

How Much and How Soon?

Let us consider synthetic protein as the first major production possibility, since rising expectations and other factors render imperative a large increase in high-quality protein in the diet of an estimated 1-billion presently malnourished people. A.T. McPherson proposed in 1973 that enough synthetic protein could be produced to bring the total amount of animal protein and its equivalent in synthetic

Table 9. World Production of Fossil Fuels and World Food Supply.

Item	Quantity (10^6 metric tons) Coal equivalent	Actual	Energy Value*** (10^{12} Calories)
Fossil fuel production:*			
Coal and lignite	2,330	2,330	16,030
Crude petroleum	2,741	2,109	18,858
Natural gas	1,307	667	8,992
Totals	6,378	5,106	43,880
Food consumption:**			
Protein		89.8	404
Fat		77.9	724
Carbohydrate		541.8	2,256
Totals		709.5	3,384
Food: Fossil fuel (percent)		14	8

*Source: Statistical Office of the United Nations (1971) World Energy Supplies. 1966-1969. Statistical Papers Series J No. 14. United Nations, New York, 10017.

**Source: Estimated from data in Economic Research Service (1964) World Food Budget 1970. U.S. Dept. of Agriculture, Washington, D.C., 20250.

***Source: Conversion factors used: Protein, lg. = 4.5 Calories; Fat, lg. = 9.3 Calories; Carbohydrate, lg. = 4.165 Calories; Coal equivalent, 1 metric ton 6.88 x 10^6 Calories.

protein up to 60 grams per day per person. Sixty grams is proposed on the basis that the consumption of animal protein in the economically advanced countries is from about 50 to 70 grams. A total of 95 grams from all sources is suggested, since this is an average amount for the developed countries. The total amount of synthetic protein to be produced for the 2,515 million people in the less developed countries in 1985 would be 43 million metric tons, as shown in Table 10.

The production of 43 million tons of a new synthetic product would be well within the capability of the world's industry. The quantity of raw materials required would be small in comparison with the amounts that are available and used for other purposes. Assuming that one ton of petroleum would be used for one ton of product, the petroleum required would be less than 1.5 percent of 1979 world production. About 7 million tons of nitrogen would be required in the form of synthetic ammonia. In comparison, the world demand for synthetic ammonia for fertilizer for 1980 is estimated at 51.5 million short tons or 46.7 million metric tons, expressed in terms of nitrogen (21).

In considering dietary protein, attention must be given to the supply of energy since, in the absence of sufficient calories, the body must use proteins for energy instead of for growth and maintenance. The FAO plan for 1985 calls for an increase over 1962 of about 300 calories per person per day, bringing the average up to 2450 calories. This amount is considerably below the 3000-calorie diet of the Western countries, but it probably would be adequate since it is 10 percent above the basic requirement, taking into account the smaller body size and lesser amount of food needed to maintain body temperature in a tropical environment.

Getting Started

There are two possible courses regarding the production of food by synthesis. If, as at present, synthetic production is left wholly to commercial interests, industry in the Western countries and Japan will move slowly from the present production of low-volume, high-unit-value items to items of larger volume and lower unit value that will find their market in countries that already have an abundant supply of natural foods. On this basis, judging from experience in the production of synthetic non-food products, an output sufficient to have significant impact on the world food supply could be achieved by the year 2000.

The other alternative would be a massive, accelerated program of research, development and production similar to

Table 10. Protein Supply for Developing Countries.

Source of Protein	FAO Provisional Plan* (grams/person/day)			McPherson Proposal** for 1985	
	1962	1975	1985	(grams/day)	(metric tons/year)
Plant	46	48	50	35	32
Animal	11	12	13	13	12
Synthetic	0	0	0	47	43
Total	57	60	63	95	87

*Data from graph on p. 507, vol. 2, Provisional Indicative World Plan for Agricultural Development. FAO, Rome, 1970.

**To furnish protein in the per person amounts shown in grams/day to 2,515 million people. Proposed by A.T. McPherson in unpublished paper at Symposium on "Food for People," Annual Conference of Society for Social Responsibility in Science, New Rochelle College, New York, October 5-8, 1973.

to but vastly more complex than the program that brought synthetic rubber into large-scale production during World War II. The requisites of such a program would be: 1) A political determination, at national and/or international levels, to enter into a significant program to produce synthetic foods; 2) adequate support in scientific and engineering manpower; 3) strong, central authority for conducting the program; 4) an organizational structure designed to facilitate cooperation between governments, universities, other research organizations, and industry.

How realistic is a major world effort to accelerate synthetic food production at this time? The major arguments against this happening -- those of technical feasibility, operational cost, and the likelihood of consumer acceptance -- have, as this chapter has revealed, been well answered over the past few decades. But the least tangible, yet most difficult argument against the undertaking, the insistence by leaders in the technical and scientific community that the "time for synthetic foods is not yet on hand," must still be answered.

In a world fast approaching the limits of its resources, mankind does not have the luxury of waiting for possible solutions to problems "when the time is at hand." The promise of producing a significant portion of the world's food in efficient factories, independent of the vagaries of climate and weather and open to the creative possibilities of chemistry, materials science and bio-engineering, surely is worth careful evaluation at this critical stage in Earth's history.

Acknowledgement

The author is deeply indebted to Dr. Archibald T. McPherson, retired former Associate Director of the U.S. National Bureau of Standards, for ideas and material used in this chapter. For the past twenty years Dr. McPherson has been studying, talking and writing about the possibilities for synthetic food production in the world food system. He graciously permitted me to borrow extensively from the following papers:

"Protein from Synthetic Ammonia: Routes for producing High-Quality Protein for Human Consumption." Indian Journal of Nutrition and Dietetics, Vol. 7, pp. 171-195, May 1970.

"Synthetic Foods: Their Present and Potential Contribution to the World Food Supply." Ibid. Vol. 9, pp. 285-310, September 1972.

"The Golden Age of Nutrition." Journal of the Washington Academy of Sciences, Vol. 68, pp. 121-128. 1978.

Copies of these are available by writing to:

Dr. A.T. McPherson
403 Russell Avenue-Apt. 804
Gaithersburg, Maryland 20760
U.S.A.

References

1. Brandt, D., L. Hontz and M. Mandels. 1973. Engineering aspects of the enzymatic conversion of waste cellulose to glucose. AIChE Symposium Series 69. No. 133.

2. Schubert, L. 1976. Synthetic foods -- a research overview. First Int'l Congress on Engineering and Food. (Unpublished paper, available from L. Slater).

3. Senti, H. 1976. Synthetic vitamin production. First Int'l Conference on Engineering and Food. (Unpublished paper, available from L. Slater).

4. Pauling, L. 1970. Vitamin C and the common cold. Freeman and Co., San Francisco, CA 94104.

5. Frankenfeld, J.W., Editor. 1968. Study of methods for chemical synthesis of edible fatty acids. NASA-SP-70 National Aeronautics and Space Administration. For sale by Clearinghouse for Feral Scientific and Technical Information, Springfield, VA 22151.

6. Sonntag, N.O. 1966. J. Amer. Oil Chem. Soc. 45, 1.

7. Sonntag, N.O. 1968. Ibid. 46, 6A, 8A, 20A, 29A.

8. Weiss, A.H. and Shapira, J. 1979. Hydrocarbon processing, 49, 119.

9. Weiss, A.H., O.V. Krylov and M.M. Sakharov. 1976. Synthetic carbohydrate from formaldehyde. First Int'l Congress on Engineering and Food. (Unpublished paper, available from L. Slater).

10. Spano, L.A. 1976. Cellulose hydrolysis to glucose. First Int'l Congress on Engineering and Food. (Unpublished paper, available from L. Slater).

11. Miller, S.A. 1964. Conference on nutrition in space and related waste problems. NASA-SP-70 pp. 343-351. (See ref. 5).

12. Dymsza, H.A. and G.S. Stoeusand. 1966. Activities Report, U.S. Army Natick Laboratories, 18, 119.

13. Swaminathan, M. 1972. Evaluation of protein quality, pp. 392-420 in Proceedings of the first Asian congress of nutrition, Hyderbad. Nutrition Society of India, Hyderbad, India.

14. Food and Agricultural Organization. 1970. Provisional indicative world plan for agricultural development: a synthesis and analysis of factors relevant to world, regional and national agricultural development, FAO, Rome.

15. Gutte, B. and R.B. Merrifield. 1969. J. Amer. Chem. Soc. 91: 501-502.

16. Denkewalter, R.G. and R. Hirschmann, et al. 1969. J. Amer. Chem. Soc. 91: 502-508.

17. Merrifield, R.B. 1965. Science 150: 178-185.

18. Calvin, M. 1969. Chemical evolution. Molecular evolution toward the origin of living systems on the earth and elsewhere. Oxford University Press. New York.

19. Inglett, G.E. 1975. Fabricated foods. Avi Publishing Co., Westport, Conn.

20. Slater, L.E. 1978. Good nutrition at low cost. Northeast experiment station collaborator's conference, USDA Eastern Regional Research Laboratory, Phila. PA 19118.

21. Harre, E.A. et al. 1972. Estimated world fertilizer production capacity as related to future demand, Tennessee Valley Authority, Muscle Shoals, Alabama.